# ENTANGLEMENTS
## TOMORROW'S LOVERS, FAMILIES, AND FRIENDS

SHEILA WILLIAMS, EDITOR

THE MIT PRESS
CAMBRIDGE, MASSACHUSETTS
LONDON, ENGLAND

This book was set in Dante MT Pro and PF DIN pro by The MIT Press. Printed and bound in the United States of America.

Library of Congress Cataloging-in-Publication Data

Names: Williams, Sheila, 1956- editor.
Title: Entanglements : tomorrow's lovers, families, and friends / Sheila Williams, editor.
Description: Cambridge, Massachusetts : The MIT Press, [2020] | Series: Twelve tomorrows | Summary: "Anthology of original science fiction short stories, published in conjunction with the MIT Technology Review"-- Provided by publisher.
Identifiers: LCCN 2019051424 | ISBN 9780262539258 (paperback)
Subjects: LCSH: Science fiction, American.
Classification: LCC PS648.S3 T83 2020 | DDC 813/.0876208--dc23
LC record available at https://lccn.loc.gov/2019051424

10  9  8  7  6  5  4  3  2  1

# ENTANGLEMENTS

**TWELVE TOMORROWS SERIES**

In 2011 MIT Technology Review produced an anthology of science fiction short stories, *TRSF*. Over the next years Technology Review produced three more volumes, renamed *Twelve Tomorrows*. Beginning in 2018, the MIT Press will publish an annual volume of *Twelve Tomorrows* in partnership with Technology Review.

*TRSF, 2011*

*TR Twelve Tomorrows 2013,* edited by Stephen Cass

*TR Twelve Tomorrows 2014,* edited by Bruce Sterling

*TR Twelve Tomorrows 2016,* edited by Bruce Sterling

*Twelve Tomorrows,* edited by Wade Roush, *2018*

*Entanglements: Tomorrow's Lovers, Families, and Friends,* edited by Sheila Williams, *2020*

# CONTENTS

# ACKNOWLEDGMENTS

Gratitude is due to Gideon Lichfield, editor-in-chief of MIT Technology Review, and Susan Buckley, associate acquisitions editor at the MIT Press for shepherding the Twelve Tomorrows series and for their guidance. I am especially grateful to Susan, my editor throughout the process of creating *Entanglements: Tomorrow's Lovers, Families, and Friends*. This book would not exist without her valuable input. Thanks are also due to Noah J. Springer, assistant acquisitions editor, for his help and persistence; Kathleen A. Caruso, senior editor, for her meticulous copyediting; and everyone else at the MIT Press who helped bring this anthology to fruition.

I am grateful to every author who contributed a story to this book. Their unique talents gave us ten diverse and unexpected ways of looking at tomorrow's lovers, family, and friends. Tatiana Plakhova's imaginative and innovative art, Lisa Yaszek's insightful profile, and Ken Liu's excellent translations skills were additional distinctive and welcome contributions to *Entanglements*.

My last words of appreciation are for some of the people who make up my own entanglements. I received practical advice and emotional support from my longtime friend, the author Jim Kelly. Finally, my husband David Bruce and our daughters, Irene and Juliet, lovingly allowed me to test our family bonds as I focused on this anthology for hours on end. Thank you for understanding.

# INTRODUCTION: ENTANGLEMENTS

## Sheila Williams

AN ENTANGLEMENT IS A COMPLICATED RELATIONSHIP. THIS IS TRUE WHETHER IT refers to a brother with a drug addition, a child genetically altered without her parents' knowledge, friends getting together for their "weekly bad rosé night," a casual hookup arranged by a personal avatar, or a particle whose quantum state can't be described independently of one or more other particles even when those particles are very far apart.

Science fiction explores the future, and it does that very well. The future can't be explored without also considering the effects that scientific and technological discoveries will have on all the relationships that tie us to each other. *Entanglements* is the sixth volume in the Twelve Tomorrows series and the first to have a central unifying theme. The book's ten fiction authors were asked to write tales about the emotional bonds that hold us together. They had a broad canvas. Their tales could be about families, friends, or lovers, but they all were asked to explore different ways in which these bonds would be affected by our ever-evolving knowledge of science and advances in technology.

I'm sure the relationships of the future, be they romances, platonic friendships, or family ties, will be just as loving, messy, complex, affirming, disturbing, heartbreaking, all-embracing, and fulfilling as they are today. They will be affected by our changing world. Infrastructure changes brought on by our warming planet, scientific discoveries, and technical innovations will put new stress on the forces that entangle us with each other even as they relieve some of the issues that complicate life.

The subjects of the fiction in this anthology range from genetic engineering to AI family therapy to neural webs, floating fungitecture, and the modern equivalent of a love potion. Friend groups form while people try to survive a terrifying natural disaster, children are oversupervised by well-meaning parents, and lovers attempt to resolve their differences by leaning on a therapeutic sexbot. A different sort of robot improves the life of a woman dealing with Parkinson's, and events get very interesting when a co-op of mothers attempts to raise a child.

The stories differ in tone and length. Xia Jia's novelette, which is brilliantly translated from Chinese by the Hugo Award–winning science fiction author Ken Liu, is a thriller set in a Buddhist monastery. Cadwell Turnbull's short story is a quiet look at a plant geneticist and her family coping with grief, while Annalee Newitz's short tale charms us with an often hilarious depiction of their character's interactions with friends and lovers.

With six Hugos and two Nebulas, and a large body of groundbreaking fiction, Nancy Kress is one of science fiction's most distinguished authors. She is known for her diligently researched tales about the ways in which genetic modification and other new technologies may transform humanity, and "Invisible People" is no exception. Nancy is profiled for *Entanglements* by Lisa Yaszek, professor of science fiction studies at Georgia Tech in the School of Literature, Media, and Communication.

The stunning artwork that accompanies the stories and Nancy's profile is by Tatiana Plakhova. Tatiana considers her work to be "infographic abstracts." She uses mixed media software to reveal the webs of data connections that exist among people, mathematics, and landscapes.

In addition to those already named, this book includes stories by some of science fiction's best known authors—Mary Robinette Kowal and James Patrick Kelly—as well as tales by exciting newcomers like Sam J. Miller, Suzanne Palmer, Rich Larson, and Nick Wolven. Our writers, who hail from across the globe, are here entangled with stories of intimacy in our technological future. Presented for your reading pleasure are ten thought-provoking stories, strikingly different and yet deeply connected. Enjoy these science fiction entanglements!

# 1 INVISIBLE PEOPLE

## Nancy Kress

## 1.

WHEN I RUSHED INTO THE KITCHEN, ALREADY LATE FOR WORK, JEN AND KENLY were bent over her tablet, Brady was flinging oatmeal from his high chair, and the wall screen blared the animal channel. Leopards flowed sinuously through tall grass.

"Why didn't you wake me? I have a deposition in twenty minutes!"

"I did wake you," Jen said. "Why did you go back to sleep?"

"And why is the TV on at breakfast? Kenly, you know the rule!"

"It's homework," Kenly said. "I have to write a report. Look, Daddy! Scientists made a baby leopard!"

A blob of oatmeal landed on my pants leg. "Damn it, Jen—"

"Daddy said a bad word!"

Jen said, "Tom." That's all she had to say. In a marriage, good or bad, one word can say volumes. This word said *It's not my fault you overslept* plus *I may choose to be the stay-at-home parent, but that doesn't mean I can control a one-year-old armed with oatmeal* plus *Lighten up. Now.*

I lightened up. "Sorry. Kenly, what's your report about?"

"Leopards. See the TV?" But she didn't meet my eyes; she knew what was coming next. Jen took the oatmeal away from Brady, but not before another blob of it landed on the cast on Kenly's wrist. She or her friends had decorated it with glitter and hearts and tiny glue-on mirrors, currently a necessity among second-grade girls.

I raised my voice to be heard over Brady's howling about the loss of his blobby missiles and the shrieking of some jungle birds on the wall screen. "When is the report due?"

"Well...the outline is due today. An outline is when—"

"I know what an outline is, honey. When were you supposed to start it?"

"Monday." Two days ago. Kenly never lies to me. And she knows I'm never really angry with her. Jen and I waited too long for her, struggled too hard, made too many sacrifices in order to adopt her. And Kenly is everything parents hope for: kind, honest, smart, sunny. All children, adopted or biological, are lotteries, and with Kenly we won big. Then we got lucky again: after all our years of failed IVF, Jen got pregnant "spontaneously" with Brady.

I said, "I'll help you write from your outline tonight."

Kenly knew a victory when she saw one and, like any good lawyer, she pushed for more. "If Mom would let me *talk* in the report like normal people, with spell-check and everything—"

"No, your mother's right. You need to learn to write and spell."

"Sophie's mom lets her use spell-check!"

Jen said, "Your mom is not Sophie's mom. And I don't know what that teacher is thinking." She walked with Brady, whose eyes drooped from the exertion of the Great Oatmeal War.

I kissed her and grabbed my briefcase, now really late for my deposition. The doorbell rang.

"Two people on the front porch," the house system said. "No matches in facial recognition deebee." The wall screen had replaced the exotic jungle birds with two strangers holding up badges.

You never anticipate the moment your luck runs out.


**FBI SPECIAL AGENTS ROSA MORALES AND MIA FRIEDMAN GAZED AROUND OUR LIVING** room, missing nothing. Not the shelf of lopsided, handmade gifts from Jen's former first-grade students. Not the three-foot-high toy space station that I'd put together wrong. Not the one expensive object, an Eric Hess sculpture that had been a gift from a client for whom I'd won a tough custody case. The object, shelved high to be safe from the kids, was spectacularly out of place. Jen wanted me to sell it, not so much for the money but because we aren't the kind of people who have museum-quality art. We aren't rich; we aren't socially prominent; we aren't saints or sinners. I don't handle the kind of divorce cases that make the news. We're invisible people, with no reason to have FBI agents sitting on our sofa, which, I now saw, had peanut butter smeared on one worn arm.

No one said anything until Jen got Kenly on the school bus and Brady in his crib. Then I said, "What's this about?"

Agent Friedman, older and clearly in charge, said, "This is about your adoption of Kenly Sarah seven years ago."

Instantly Jen went on the attack, a lioness with cubs. "There shouldn't be any problem with that. We have legal adoption papers, we went through proper channels—"

"Yes," Agent Friedman said, "but unfortunately, the adoption agency did not."

"What's that supposed to mean? My husband's a lawyer and—"

"Please calm down, Ms. Linton. Neither you nor your husband did anything wrong, and the child is legally yours. We're here to ask you exactly how the adoption progressed. The Loving Home Adoption Agency may be involved in violations of U.S. Code Title 18, Chapter 96."

I said incredulously. "The RICO Act? Racketeering?"

"Engaging in a criminal enterprise, yes."

"How?" Jen said. "Kenly wasn't bought illegally or anything. We met with the biological mother once and talked to her through an interpreter; she was accidentally impregnated by her boyfriend who then skipped out, and her religion forbade abortion. All we did was pay for her medical expenses and care during pregnancy, and we were in the hospital when Kenly was born! St. Mary's Hospital!"

"Yes, we know," Agent Friedman said. "And eventually you may be called on to testify about all that in court. But for now, we just want to hear what happened from your perspective."

Jen said, "And no one is going to try to take Kenly away from us?"

"No, ma'am. I can promise you that."

It was what Jen needed to hear. The lioness morphed back into my wife. I said, "I want our lawyer present." I am an attorney, but a divorce lawyer is a long way from racketeering indictments.

"If you wish. Meanwhile, I can at least tell you that your daughter is not the result of an accidental pregnancy, as you were told. She is the result of an offshore operation that hires indigent women to carry IVF embryos to term in order to be adopted out. You've had annual follow-up visits to the Loving Care Agency, right? Visits that included interviews with both of you, a well-child medical exam of Kenly, and a detailed questionnaire?"

Jen said, "That's all part of our contract with Loving Home. That we participate in a long-term study of adoptee adjustment."

"Not exactly," Agent Friedman said.

Brady began to fuss. The robotic arm on his crib activated and checked his diaper, then dangled a toy in front of him. He went on fussing, but for the first time ever, Jen ignored him. She demanded, "What did they do to Kenly? In those medical exams? I was right there and—"

"Nothing in the medical exams. It happened long before that, during in vitro fertilization." Agent Friedman hesitated, then apparently made a decision to say more—maybe because I was a lawyer and would find out anyway, maybe because we looked conventional enough to be trusted, maybe even out of sympathy. She said, "Your daughter's genes were illegally altered. Illegally and without consent."

"Altered? How? She's a normal seven-year-old, healthy, nothing different about her—I don't believe you!"

"It's the truth. I'm sorry."

"Have other kids been 'altered'? Who are they?"

"The FBI cannot give out names of other potential witnesses."

"You didn't answer my first question! Altered *how*?"

Brady went from fussing to full-out howl. Jen didn't move. Neither did I.

Agent Morales spoke for the first time. Her coloring matched Kenly's: smooth tan skin, exuberant dark curls, deep brown eyes. There was even a faint island lilt to her voice.

She said, "How did Kenly break her arm?"

**LEOPARDS**
**By Kenly Linton**

Some syentists made a Amur leopard. That is one kind of leopard. It went
xtink many years ago. The syentists found its genes someplace and
put them into a African leopard and the baby was borned! It is very cute.
The mother licks it. That is leopard kisses.

**2.**

MARY, MY ASSISTANT, HAD RESCHEDULED THE DEPOSITION, BUT I HAD A NEW CLIENT coming in at eleven o'clock. Until then, I sat in my office with the door closed, a cup of coffee growing cold beside me, and stared at the picture of my family on my desk.

Jen, laughing, her hair blowing in an ocean breeze.

Agent Friedman said that the scientists who "altered" the genes in the embryo that would become Kenly—those unidentified people—were part of a large, well-funded, offshore private organization. They implanted the embryos in poor young single girls who desperately needed the money, and then adopted out the babies through agencies like Loving Home. The girls were paid only if they agreed to parrot the pregnancy story they were given.

Brady, six months old, grinning around his first tooth.

The FBI would not tell us the name or location or purpose of the organization because it was "part of an on-going investigation." But it seemed to be an exercise in eugenics, that disgraced twentieth-century idea, done with twenty-first-century genetics.

Kenly in a ruffled blue swimsuit, pointing proudly to the sandcastle she'd just built, pail-shaped and topped with a seagull feather.

To be told even as much as we were, Jen and I had to sign papers swearing us to silence until the case came to trial.

Behind my family, the vacation cottage we'd rented on Maryland's Eastern Shore, gray clapboards weathered by wind and wind-borne sand. Every year we rented the same cottage for two weeks.

Agent Morales had described the knockout technique for cutting out one gene and replacing it with another allele of the same gene. Apparently it had been used for decades on mice, in research, then on other animals. Recently something called "Curtis tools" had made a huge leap in gene-altering precision and safety. In the United States, it was illegal to genetically engineer human embryos to birth.

Every year, as we packed up to leave the Maryland cottage and drive home to North Carolina, Jen and I discussed buying the cottage. Every year we decided against it. The failed IVF attempts, the fees to the Loving Home Adoption Agency, the kids' college funds. My modest four-partner firm didn't get the high-profile cases. Jen and I couldn't afford the cottage.

Jen shouting at the FBI agents: "I want to know what was done to Kenly! Why are you here if you won't tell us anything specific?"

Agent Morales said, "We're here for your protection, and your daughter's. The genes that have been altered involve increased risk taking. You need to watch Kenly especially carefully. That's all we can tell you."

It made no sense. Kenly was not a risk-taker. Unlike other children we knew, she didn't climb trees or her swing set or, like Bobby Cassells, the porch roof. She didn't swim out farther than we allowed her. She didn't race her bike down steep hills, unlike her friend Sophie Scuderi, who last month was taken to the emergency room with facial lacerations. Kenly had broken her arm in the most mundane of ways: tripping over a tree root in a neighbor's backyard. What was "increased risk taking" about that?

And why would an organization—any organization—go to the expense and danger of altering risk-taking genes in a few random children?

We wouldn't be going to the Maryland cottage this August. It had been destroyed by Hurricane Lester, one of the many major storms we'd all come to accept as normal as climate change worsened.

Mary poked her head into my office. "Tom, your eleven o'clock is here."

"Send her in. And after she leaves, get George in here. Tell him—her—damn it, *them*—that it's urgent."

Mary looked startled as she withdrew her head. I don't swear in the office, and usually I remember George's recent switch of pronouns.

The moment the client walked in, I smelled money. She wore an expensive summer suit—a divorce attorney has to develop knowledge of class markers including women's fashion, a subject more complicated than torts. Amanda Wells Bryant had the perfectly coiffed blond bob of her tribe, a successful facelift, and discreet bling. She also had the sourest expression I'd seen in a long time.

"I want to divorce my husband," she said, in the tone of one used to ordering around luckless servants. "And I want as much of our money and property as possible. Preferably, all of it."

"I see," I said. "Tell me—"

"You don't see," she said. "Our finances are very complicated. They include homes in France and St. Barts, and multiple companies and leveraged holdings. But I'm told you're a good divorce lawyer and your team won't gouge me with your fees."

Rich *and* cheap. That's why she chose me instead of a white-shoe law firm. Some of my clients need hand-holding, some need instructions on basic financial instruments, a few actually want a fair division of assets. She wasn't any of those types.

"Tell me your story, from the beginning." I already knew it from just looking at her, but it turned out I was only partly right. There was another woman, of course, younger and fresher. Amanda had already hired a private investigator and had compromising photos. The surprise was the speed with which she wanted the divorce to happen.

"I want it all concluded before the bastard is done with his deployment."

"He's in the military?"

"He's commander of a nuclear submarine at sea somewhere in the Arctic, a radio-silence tour of duty. Those usually get extended from three months to five, given the situation up there. I want the fucker to come home to nothing, locked out of all our houses, as penniless as you can make him. I'd like to ruin her, too, but I suppose that isn't possible."

"Afraid not." I've had vindictive clients before, but she chilled me as much as the icy welcome she wanted for a man risking his life in the Arctic. Tensions between Russia, Canada, and the United States over the newly ice-free Northwest Passage could escalate at any moment into a shooting war.

She said, "With any luck, his sub will be torpedoed and I won't need you at all."

GEORGE WHELAN IS THE BEST INVESTIGATOR I HAVE EVER WORKED WITH. I HAVE others to do the tedious, painstaking computer investigation that so often provides George with financial leads, but for on-the-ground legwork, no one beats George.

Until six months ago, they were Georgiana, named for a distant ancestor who was actually a notorious British duchess. Now gender-fluid and with a new pronoun, George could blend in anywhere as male, female, or androgen, conventional-looking or flamboyant, teenaged rocker or thirty-something businesswoman. When they want to, they could also be invisible. I have passed George on the street without recognizing them. The three other attorneys in my firm envy me George, who can find anything, anywhere. They are expensive and worth it.

Today George wore jeans, a faded blue hoodie, a man bun, no detectable makeup. Actually twenty-nine, they could have been nineteen, somebody's nephew visiting the office. "What's up, Tom?"

"Something off the firm payroll. For me personally."

Blunt as always, George said, "Can you afford me?"

"Yes. Usual hourly rate plus usual expenses." It would be a stretch, but the vacation cottage was gone anyway.

"This isn't . . . I can't believe . . . Is Jen cheating on you?"

"Cheating? God, no! George, the FBI came to our house and I'm not supposed to tell anyone what they said. But I'm going to tell you. You okay with that?"

"Sure. Are you? You'll be the one breaking the law, taking on the Fibbies."

"Only if they find out. I need to investigate the Loving Home Adoption Agency, find a girl somewhere in Raleigh-Durham who gave up her baby to the agency seven years ago, and, from what you can get her to tell you, determine where her pregnancy actually came from. The child is Kenly Sarah Linton, my daughter. I have some names and dates, but not many. Are you in?"

"Absolutely. Tell me everything, every small detail, from the beginning."

My usual phrasing, but this time I was on the other side of the interrogation. George recorded me. It was the only time I ever, in all our investigations, saw their eyes widen. When I finished, they said simply, "Why?"

"That's part of what I want you to find out. Who, where, and the fucking why."

> **There are diffrent kinds of leopards like snow leopards and clowded leopards and African leopards. Leopards eat other animals. Once a African leopard killed a babune to eat it. But the babune had a new tiny baby! The leopard didnt kill the baby babune. It put it in a tree to save it from higheenas. It licked the baby babune. You can see the video on line. Look it up!**

## 3.

WHEN I GOT HOME, JESSICA, OUR TEENAGED OCCASIONAL BABYSITTER, WAS CHANGING Brady, Kenly was watching TV—usually forbidden on weekdays—and Jen was at the computer in my study, still in the bathrobe she'd worn that morning. She scowled at me.

"It's so confusing! Some studies say that genes for dopamine receptors like DRD4 influence risk taking, especially if you have seven repeats of the gene. Other studies say no, it isn't dopamine, it's glutamate and gamma-aminobutyric acid, neurotransmitters in the brain. Some scientists say there are more than a hundred genetic variants linked with risk taking, but even combined they account for only about 2 percent of differences in risk taking among people. How the hell can they figure out that? Then more studies say none of those studies are reliable because they use self-reporting, and people lie. I don't... I can't figure out..."

She was trembling. I took her in my arms. Her hair smelled dirty. I held her closer. Jen and I do this for each other: switch roles from comforter to one who needs comforting. I see a lot of broken marriages, and I know how good it is that we aren't each locked into one role.

When she stopped trembling, I said, "Tell me everything from the beginning, every little detail."

She pulled away and smiled wanly. "You want to take a deposition?"

"Yes. You want a lawyer present?"

"Fortunately, I have one."

We talked for a long time. She told me about the research she'd found on risk taking, which was confusing, although presumably not to scientists. I told her about George, and how much they were going to cost us.

Jessica knocked on the door. "Kenly wants to go to the park. Is that okay? Brady's asleep."

"Sure," Jen said.

We resumed our conversation, minutely examining Kenly's behavior for all of her seven years, comparing it to other children's, and ending up as baffled as before. "I want to have a full gene scan done on Kenly," Jen said. "Not the kind that just tells you where your ancestors are from—the full real thing. So I can compare it online to that of a normal seven-year-old girl."

"Honey, I don't think there's such a thing as 'normal.' The alleles—"

"You know what I mean! Don't nitpick!"

She was a tinderbox, and I was not going to light a match. "Yes, I know. We'll do it."

"I'll find some place and make an appointment for tomorrow, I—"

The kitchen door slammed and Jessica's voice, uncharacteristically loud, said, "Don't you ever do that again!"

Jen and I raced to the kitchen. Kenly stood with her purple backpack at her feet, and Jessica—Jessica!, eighteen, mathlete, Jane Austen lover—held a Glock subcompact handgun.

Jen grabbed Kenly by the shoulders. "Are you all right?"

"Yes. I hate Jessica!"

"No, you don't. Jessica, what *happened?*"

I said, "Are you licensed to carry that weapon? And did you fire it?" *Please, God, let her say no.* But my mind raced through names of criminal defense lawyers, state gun laws, and bail bonds.

Jessica, pale but coherent, said, "I'm licensed for conceal carry in West Virginia, and North Carolina honors all other states' permits. I fired the gun into the air just to scare him off."

"Scare who off? Jessica, can you tell me everything from the beginning, every small detail? Sit here, at the table. You're not in trouble, we just need to know." I used my most reassuring voice, but I'm not sure Jessica needed it. She was clearer and more thorough than 90 percent of the adults I put on the witness stand.

When Kenly's TV program had ended, she started her math homework on the coffee table. Jessica had put on the news, part of some homework assignment of her own. When Kenly's multiplication worksheet was done, she wanted to go to the park. She'd gone upstairs first and put on her backpack, which Jessica thought was odd but harmless. The backpack was new, purple with tiny mirrors sewn in a flower pattern, and Kenly loved it. They'd walked to the playground, and then abruptly Kenly had run from Jessica and wouldn't return. Jessica ran after her, but Kenly was fast and Jessica, overweight and no athlete, was not. Kenly had run straight into the homeless camp at the edge of the park.

"How did she know the—"

"It was on the news. She must have been listening. They said there were children there, and they didn't have toys."

Kenly started emptying her backpack of American Girl dolls, stuffed animals, and last Christmas's prize toy, Astronaut Jane Genuine Flight Control Console, $96.99 if you could find it at all. She called out for kids to come get toys. No kids appeared. But two men were there and one of them grabbed Kenly and said, "What else you got, girlie? Money?" The other said, "Let her go, Sam, you're drunk," but the man started to reach into the pockets of Kenly's jeans.

"That's when I fired into the air," Jessica said. "I think the other guy would have made him let her go, but I wasn't taking any chances."

A siren sounded, distant but coming rapidly closer.

All at once the self-assured junior superhero vanished and Jessica looked scared. "Will they arrest me?"

"No," I said. "I'll talk to them. And you will, too, exactly as you told me. It'll be okay, I promise."

"And you," Jen said to Kenly, "if you ever do anything like that again, we'll—"

Then Kenly shocked us more than Jessica's gun, more than Kenly's mad flight toward the homeless camp, more than the man in the park. Kenly stamped her foot and glared at us all. "I *will* do it again. Those are kids with no toys, not even one tiny damn mirror."

She burst into tears.

I DIDN'T EXPECT TO HEAR ANYTHING FROM THE FBI, AND I DIDN'T. RICO INVESTIGATIONS can take years. I did expect to hear from George, but all I got for a month was a staggering expense bill. I wired the money to Miami.

A week later, I wired more funds to Georgia.

I was going to need a big partial payment from Amanda Bryant, whose commander husband must still be alive on a submarine gone quiet, since the navy had not notified her otherwise. Each day the situation in the Northwest Passage got worse. The United States sent warships, the Russians sent warships, Canada filed protest after protest, the ice continued to melt. So far the shipping lanes were still open, the warships' guns silent. So far.

"I hope his sub is sunk," Amanda said. "And did you find documentation for that Ukrainian shell company I told you about?"

"Not yet. We're looking."

"Well, find it. I want that money before we go to court!"

"Amanda, I've told you that Commander Bryant will be entitled to time for his lawyer to prepare his side of the case."

"There is no 'his' side. He can have his whore. I get everything else."

Had she once loved him? Had they ever laughed together, touched each other fondly, shared daily news over cups of coffee? Hard to believe.

"You better win this case," she said.

"I will." It was my job, my oath as an officer of the court, to represent my client's best interests. Even when my client was a bitch.

Kenly had turned sulky with us but otherwise behaved as usual. Not that she had much chance to do otherwise—Jen accompanied her everywhere, even if she had to lug Brady along. Kenly's gene scan showed nothing abnormal in the markers that science had already decoded. Some of the genes tentatively identified for risk taking were present, some weren't. "But you understand," the genetic counselor said for the third time, "that we have only identified proteins made, diseases caused, and genes cross-interacting for a small percentage of the codons. Genomics is in its infancy, but it's evolving rapidly. There are groups working on genomics at universities, pharmaceutical companies, government laboratories."

Jen and I glanced at each other. The glance said, *... and an illegal lab of unknown purpose in an unknown country doing unknown things to human embryos.*

In the car on the way home, after a long silence, Jen said, "They must at least be a hell of scientific group. To do that and still produce healthy kids. Or else the FBI is just wrong about the whole thing. Or lying to us."

"Why would—"

"I don't know. I don't know anything."

**Leopards know a lot. They know how to hunt and have baby leopards and gard their teratory. Sometimes leopard mothers will take care of babies that are not even theres. I think leopards are good but**

I wish they didn't eat babunes. When I grow up I will be a animal trainer and train leopards to eat something else and be nice to babunes. Maybe nuts and berrys which are helthy anyway. Or bananas, like babunes do. Or leafs.

## 4.

GEORGE, LIKE KENLY'S LEOPARDS, KNEW A LOT.

They strolled into my office, this time as Georgiana, with lipstick, earrings, and a maxiskirt with combat boots. I told Mary to postpone whatever was on my calendar for the rest of the afternoon.

"Good," George said. "I've got a lot to tell you. And a surprise."

"The surprise first."

"No, it's arriving separately. Tom, this is a big operation. But first, I have to tell you that the FBI caught me snooping, grilled me for a few days, and forbid me to poke around anymore or they'll charge me with interfering with an ongoing investigation, obstruction of justice, and anything else they can make stick. So I'm off the case, but you won't need me after the surprise arrives."

"Fuck it, George—"

"Georgiana."

"—I don't want any games! Just tell me what you found!"

If George was startled at my uncharacteristic tone, they didn't show it. "Okay. I found Kenly's biological mother and made her talk. It didn't take much, just mild intimidation, since these girls—and yes, I found more and I'm coming to that—are poor and vulnerable. They're all in the country illegally and terrified of being deported. The money you paid Jimena was supporting her entire family in the Dominican Republic. She birthed another genetically engineered baby after Kenly, and she's now pregnant with a third."

"The group doing the engineering is based in the Caymans. I found it, which is how the Fibbies found me. I never got inside—let me tell you, the Pentagon doesn't have security as good as this place. But my team photographed and checked out everybody going in, and it's an impressive list of scientists from four different countries, all with sterling reputations in genetics. I'll give you the list. A lot of truck activity. Some go to the airport, and then biological coolers are hand-carried aboard planes going all over the Southeast United States, France, England, and China. Scientists in the organization, which seems to be nameless, are from those countries. They—"

"But what are they doing to the embryos? And why?"

"I don't know yet. Wait for the surprise to arrive in"—George checked their watch—"about twenty minutes. And let me finish. This organization is *big*, and that takes big money. It's filtered through so many shell companies and Swiss accounts that I don't think even the FBI is going to be able to trace it. Although maybe they can—they have resources that my team doesn't. The point here is that this group is furiously altering embryos, impregnating young girls, taking excellent care of the surrogates, and adopting out the infants through legitimate adoption agencies that, so far as I can determine, might not be aware of the source of the pregnancies. Or maybe some know. The Caymans organization must have excellent cybersecurity because my guys couldn't hack their records, which frustrated the hell out of them. One actually threw his monitor against a wall."

"Will you please—"

"Seventeen minutes, Tom. Here's a major point—if these genetic alterations are dominant, Kenly's changes will be passed on to her kids. If the alterations include something called a gene drive, which I only learned about on this investigation, the altered genes get passed on to even more of her descendants than they would be ordinarily. This group, this pack of internationally distinguished scientists, is trying to slowly change the human race."

"To take more risks? Why? We already take too many risks with the future!"

George stared blankly. *Too many risks* was a foreign language; George assumed risk like a fish assumed water. I wasn't about to lecture them on the dangerous standoff in the Northwest Passage, the divorces I saw caused by stupid and chancy drug use, the carbon emissions that risked coming generations' future. I was too angry.

"If you don't tell me what this 'surprise' is—"

"I think it's me," a voice said.

She strode through my office door, Mary sputtering ineffectively behind her like a dory in a warship's wake. With effort, I kept my jaw in place. Kathleen McGuire was instantly recognizable from the news, any news. The heir to oil and shipping money, she'd then founded an investment firm that specialized in financial instruments as complicated as astrophysics. In her sixties, she'd never had work done and her face, although lined, was still beautiful. The huge blue eyes and red hair—surely, by now, dyed—were only part of it, as was her perfectly tailored suit. Couture, I

13

guessed, but this wasn't clothing I ever saw in my office. She made Amanda Bryant look like a middle-school teacher.

"You're Tom Linton," she said. "You're the one whose nanny fired a gun, which attracted press attention. Fortunately, it went no farther. Don't let anything like that happen again."

"You can't just—"

She ignored me. "This rogue organization made a mistake. One of those genemod kids went to my niece Valerie. Your excellent investigator here found Valerie, and so me. Your daughter Kenly is another victim? There are four of us then, parents and relatives George found that received altered babies. We will band together to bring this group down. But first we need more information."

"Yes, do you—"

"George was unable to find out just what genes have been altered, and of course you already know that there's no reliable 'standard' reference genome, but there's another way. With enough computing power, which I will hire, we can have the genomes of all affected children compared to each other, to see where alleles match to a confidence level sufficiently beyond chance. Then we can have scientists examine the literature to find studies showing these genes have identified proteins that influence identified behavior. From there we can build a legal case. The key is finding more of these kids and persuading their parents to cooperate."

I said, "The FBI—"

"Can squawk and threaten all they want. Nothing we're doing is illegal, and my lawyers are not impressed with threats. You're a lawyer, Mr. Linton?"

"Yes, a divorce lawyer, and—"

"That's not much use to us, but your cooperation is. George will remain behind the scenes to coordinate the investigators I hire to find more parents. Whoever is trying to play God with our kids will be brought down after we have enough evidence."

"Why would anyone, especially a group of 'distinguished scientists,' want to increase risk—"

"I don't know. That's what we'll find out. But we need information, starting with your daughter. I've given George the questions I want answered, but let me start. Is Kenly physically healthy?"

"Yes, but she—"

"No major diseases since birth?"

"No, and—"

"How old is she?"

Kathleen McGuire was a force majeure. Even George had not corrected her use of their pronouns, or their current name. I was determined to at least have some active part in this discussion. I said loudly, "Kenly is seven, almost eight. How old is your niece's child?"

"James. He would have been six. He's dead."

JEN BEGAN HOMESCHOOLING KENLY. KENLY HATED IT. SHE KNEW WE WERE MAKING sure that one of us was with her every minute but she didn't know why, and we couldn't tell her.

"I want to see my friends!"

"They can come over to play after school."

"I want to go to school! I'm missing stuff!"

"Your mother is a certified teacher, Kenly. You're not missing any schoolwork."

"She can't teach me gym! Or art! Not like Ms. Lentini did!"

Our sunny, cooperative little girl turned sullen and dour. The weather turned rainy for weeks; low-lying areas of the city flooded. People were rescued from rooftops by helicopters, from second-story windows by boats. A Good Samaritan drowned trying to save a woman swept away in a flash flood. In the Northwest Passage, a shoulder-launched missile was fired from the Canadian shore at a Russian warship and the world held its breath, but it was inconclusive who had fired the missile, which missed the ship. The Russian vessel didn't return fire.

I had a new divorce client, a tall thin man, wispy as a reed, and, I first thought, just as pliable. His wife of eighteen years, who'd left him, wanted the house. "Let her have it," he said. "I don't want it."

"Are you sure? It's the major marital asset and I advise that—"

"Let her have it."

I looked at him more closely, and revised my first judgment. This wasn't passivity or generosity. The reed was a toxic plant. I said, "Why?"

"She don't know this, but a big company bought twenty acres next door. They're gonna put in a wind farm. Those whirling things and the noise they make will drive her crazy, and the value of that farmhouse will drop like cement. There's a NIMBY group fighting it, but they're gonna lose. Cora don't never pay no attention to anything but herself, so she don't know about any of it. Let her struggle with the windmills the way I struggled to support her all these years while she sat on her ass and barely even cooked for me."

I said, "Her lawyer will find out about the proposed land use for windmills."

"She don't got a lawyer. Too cheap. Just make up the papers and get her to sign them fast."

My job is to represent my clients, not to like them.

Neither George nor Kathleen McGuire sent me reports about their investigation. Instead, a young woman who looked fifteen, but was actually twenty-seven (I asked), showed up every few weeks to talk confidentially to Jen and me. We three sat around the kitchen table after the kids were in bed, glasses of wine on Jen's lemon-patterned placemats, the whole scene so normal that I sometimes got vertigo from the contrast with what the young woman told us.

Two more parents of the gene-altered kids from "the operation" had been located. Then another one, then three more. Kathleen's scientist wanted ten complete genomes to run matches on. Two of the parents refused to cooperate. One didn't believe any of this had happened ("My kid's normal! Go away!") The other believed it but was too afraid of "the authorities" to want to participate.

Two more were found. Then three more. George had always been really good, and apparently so were their investigators.

Early on, I googled Kathleen McGuire's family. I found her niece's child's funeral notice. James Niarchos Carter, aged six: "Suddenly." Private funeral, donations in his name to St. Jude's Hospital, no flowers. Also no details, nowhere on the Internet. If there was a police report of an accident involving little James, it had been scrubbed from public records. Could Kathleen have that done? I had no idea how much her power and money could do.

When George had found their ten kids, creating the children's complete genomes and comparing them to each other could begin. But if it yielded matches in some alleles, the scientists would then have to figure out what those genes did. And then what? From where Jen and I stood, invisible on the sidelines, it seemed a hopeless task. We didn't see how it would help Kenly.

But it was all we had. That, plus the FBI investigation, plus trying to keep Kenly from doing anything risky. We could do that now; she was seven. What about when she was sixteen? Or twenty-six?

I didn't want to think about that.

One reason leopards are a little bit xtinct is pochers. They are terrible peeple who kill leopards to make rugs. If I saw a leopard rug I would tair it into little peeces. Pochers kill other animals too like elephants. Who wood do that? It is terrible terrible terrible. If I saw a pocher I would shoot him dead.

This is the end of my report. It is the longest report I ever rote.

**17**

## 5.

"DADDY, CAN WE PLEASE GO TO THE PARK? IT'S SATURDAY AND SOPHIE OR OLIVIA can't come over to play and it's so sunny out!"

Kenly stood by my desk, which was piled with work I'd brought home to do over the weekend. Jen and Brady, who both had colds, were napping. I didn't want to go, but Kenly looked so pathetic, a small prisoner in her own home. "Sure, Kennybug. Let's go."

Spring filled the park: tulips and daffodils and the smell of cut grass. People strolled, smiling; dogs strained at their leashes; children ran and shouted. I held Kenly's hand and she skipped along in her red sneakers. Jen had sewn the ubiquitous tiny mirrors on the back pockets of Kenly's jeans. She smiled at me, the first smile I'd had from her in weeks, and I thought my heart would burst.

"Can we get ice cream?"

"We can indeed. I want chocberrycocolimehazelnutmarshmallow."

"That's not a real ice cream!"

"Yes it is, and I'm going to have fourteen scoops of it. I'm going to—Kenly!"

It happened so fast. I'd always heard that time slows down in danger, that every moment is separate and crystal clear. This wasn't like that. One second Kenly was holding my hand and laughing, and the next she'd torn free, a running blur, the mirrors on her jeans twinkling in the sunlight. The dog that had broken its leash was a brown blur, and the toddler screaming in its jaws was noise and thrashing motion. Then Kenly was, too, pounding on the dog's head and yelling, "Let go! Let go of him!"

It did, and turned on Kenly, fastening its teeth on her leg and taking her down. Everyone was screaming, the air itself shrieked, and I was on the ground, pulling at the dog and beating it. The dog would not let go. The toddler was snatched up by somebody, but the dog still had my little girl and there was another sound, inhuman and inarticulate, and I was making it.

Then water. It hit the dog in the face and showered over me and Kenly, somehow becoming part of the noise. More water in a narrowing stream, and when the hose shot hard into the dog's face, it let go. I grabbed Kenly and ran. When I stumbled, someone grabbed both of us and set us upright.

"I've called the cops and an ambulance. I'm a park ranger. Stay right here, please."

I couldn't talk, couldn't think. In my arms Kenly, drenched and bloody, cried out. I couldn't decode the words, and then I could.

"Is the baby all right?" Kenly sobbed. "Is the baby dead?"

The same voice said, "He'll be fine. The baby is fine." Then to me: "No, sir, stay right here. The ambulance is on the way."

It was then, in the middle of the noise and blood and a stranger's calm voice that I suddenly knew what had been done to Kenly's genes.

"IT'S NOT AS BAD AS IT LOOKS," THE ER DOCTOR SAID. "YOU'RE LUCKY THE DOG WASN'T a pit bull."

*Lucky.* I was lucky. We were lucky. The dog's owner furnished proof of rabies shots and agreed to pay for all medical treatments. The mother of the toddler, not mollified, yelled at him loud enough for the whole ER to learn that she was going to sue the pants off him and make sure the dog was destroyed.

I took Kenly, drowsy from painkiller, home in a drivie cab. I couldn't let her sleep yet. I had questions.

"Kenly, that dog could have killed you. Why did you risk your life for that baby?"

She frowned. "You risked your life for me."

"You're my daughter!"

"He's my... my..."

I held my breath.

"He's a person," she finally said.

We stared at each other in mutual incomprehension. No, not mutual. I understood Kenly, but she did not understand my placing her life over all others because she is my child. She didn't understand, at a basic hard-wired and preverbal level, the kin-based allegiance that had, through all of human history, been an evolutionary force to aid survival. All my research said that genes were selfish. Sacrificing self for kin was one way that genes survived, with the greatest sacrifices for those who shared the most genes. Hadn't some famous scientist joked that he'd gladly die for two brothers or eight cousins?

But not everyone. A man loses his life trying to save a stranger from floodwaters. A soldier throws himself on a grenade to save his platoon. A philanthropist donates large portions of his fortune to cancer research, or humanitarian aid to some drought-ravaged nation he will never visit, or a secret organization in the Caymans. And Kenly risks her life to pull a toddler from a dog's jaws. She breaks her arm tripping over a root in her hurry to help a baby bird that had fallen out of its nest. She tries to give her toys to homeless children.

Survival of the fittest was not the only evolutionary force that had aided human survival. The other one had been controversial for a very long time, all the way back to Darwin.

After Kenly was safely asleep in her bed and I'd told Jen everything, I called Kathleen McGuire, giving her phalanx of assistants the code she'd designated for immediate and unquestioned access. The code worked.

"Ms. McGuire, the gene comparison might not show any alteration in genes associated with risk taking."

"They don't show alterations," she said. "I just got the genomic comparison data. How did you know?"

"Because the risk taking is collateral. But there *are* genes in the data that seem to be altered in the same way for all the kids, right?"

"Yes. Five of them. But my scientists say it's not known what proteins they code for, or how they interact, or how they affect behavior."

"Kathleen—how did your niece's son James die?"

Her voice could have re-frozen glaciers. "I'm not going to discuss that."

"Okay, but it's relevant. He died trying to help someone else, didn't he—some other kid or animal. No, don't interrupt me. I know what those altered genes do.

"They code for behavior to aid survival of the human race, even at the expense of the individual. They code for altruism."

THE SPRING AND SUMMER PASSED. KENLY'S LEG HEALED. WE DIDN'T LET HER LEAVE the house alone. She wheedled and begged and cried and guilt-consumed Jen and me, but we held fast.

George's investigative group found twenty-five more gene-altered children, tracing them through surrogate mothers and horrified adoption agencies. As more people realized something unusual was going on, the press began sniffing around, but so far no reporter had enough information to break the story.

The Arctic Council, backed by the United Nations, finally decreed that Canada had jurisdiction over the Northwest Passage. Canada ordered the Russian and American warships out, but pledged that all nations could use the passage for commercial shipping but not for military activities. For a day it looked as if the warships might not leave, and the world braced for nuclear war. Then both countries pulled out. Amanda Bryant's commander husband finished his submarine tour, came home to his mistress, and was served with the divorce papers I'd prepared. He promptly hired a lawyer. The case was thus guaranteed to drag on for a long time, furiously for the litigants and lucratively for me.

Lucas Wibberly's divorce was settled quickly. His selfishness paid off; she got nothing but the farmhouse. The NIMBY group failed to block construction of the wind farm, and the ex-Mrs. Wibberly was stuck with a house she didn't want to live in and couldn't find a buyer for.

The newly elected U.S. president removed all the previous administration's caps on carbon emissions, and global warming continued. Low countries flooded, average temperatures edged up another notch, severe storms increased, tropical insect–born diseases moved farther north. Corporate profits rose.

It was a hard summer for Jen and me. Kenly turned more and more defiant under her protective house arrest. I found it harder to litigate for clients I could not respect. Jen's cold turned into pneumonia, which meant hiring a live-in nanny to care for the kids until Jen was no longer infectious. The only bright spot was that while her mother was ill, Kenly lost her sullenness and helped with Brady and with simple housework.

Then, in August, there was another bright spot. One night, just as we were going to bed, George came to the house, smiling.

SCIENCE DOESN'T PROCEED IN STRAIGHT LINES. GREGOR MENDEL DISCOVERS THE laws of inheritance and ninety years pass until Watson and Crick put a shape to genetic structure. It's sixty more years until the first mostly reliable gene editing tool, and then ever shortening time jumps as techniques leap forward in precision and scope. Now, with major advances every few years, we can alter genes so much more than we ever thought we could, and so much more than laws allow.

But laws, too, undergo punctuated evolution: periods of inertia are followed by periods of quick change. In the United States, it was still illegal to alter human embryos. It was not illegal to develop gene therapies—genetic changes inserted into the human cells of children and adults via viral vectors—that could combat gene-caused diseases like cystic fibrosis and hemophilia.

"And also combat what the organization in the Caymans did," George said. They were hypermasculine at the moment: flannel shirt, jeans, work boots, mustache. Was the mustache fake? I had no idea.

Jen said, "Kenly doesn't have a disease."

"There's a gene therapy being tested at Berkeley that is adaptable to the kids' conditions. The vector to deliver the new genes is delivered by liposomes, which is safer than using a virus. The researchers there are eager to see if it works. And Kathleen got a compassionate use exception to full FDA trials."

Jen said, "They want to experiment on these kids!"

I said, "Compassionate use exceptions are for people who are dying."

"Then Kathleen got an exception to the exception," George said impatiently. "She has a lot of influence. Guys, this is a way to reverse what was done to Kenly."

"An untested way!" Jen said. "Kenly is not some lab rat!"

George said, "It's not completely untested, and not just on animals. One parent already had it done on their four-year-old, and he's fine."

I said, "I want to talk to that parent."

"You can't. Anonymity was part of his deal. The press is going to get this story soon, and nobody wants their child splashed all over the internet. Also, although I don't have confirmation of this, my sources say the FBI is close to indictments, which may stop the Berkeley group from proceeding with their experiment. You need to decide now."

Jen said, "Don't pressure us!"

I put a hand on her arm. Unlike me, Jen is not used to getting clients, witnesses, and juries to cooperate. I said, "George, don't think we're not grateful to you for all you've done. You and Kathleen. A chance to reverse what was done to Kenly is more than we expected. You've done a phenomenal investigative job. It's just a lot to take in, and we need a little time."

"Don't pull your lawyer tricks on me," George said, but they smiled. "You don't have a lot of time. Here's the home phone number on an encrypted line for the lead scientists at Berkeley. She says call anytime as long as it's soon—she wants your decision. I gotta go see some other people."

After they left, Jen said, "No. We can manage without some experiment on Kenly. Don't you know the history of scientists experimenting on people? Tuskegee with syphilis, Crownsville with drilling into brains, Sloan-Kettering with cancer cells injected into—"

"Stop. I know. I've done the same research you have. We have to talk this out completely, Jen. From the beginning, every small detail."

"Don't you dare treat me like you're taking a deposition!"

I apologized. Jen apologized. Then we sat in our living room, close together on the couch, and talked as a sliver of moon rose beyond the window, no bigger than a child's fingernail. Moonlight glinted on the edges of our wineglasses like sunshine on Kenly's tiny mirrors.

The treatment was experimental.

Risk taking was part of altruism, and two of the altered children had already died taking risks.

Genes were complicated things, and you don't just charge in and alter them without the risk—that again!—of turning on or off other genes. When Kenly finished the therapy, would she still be Kenly?

Would she still want to help people, to give selflessly? There was so much selfishness in the world. I saw it every day in divorce cases. I saw it on the news, whole countries risking nuclear annihilation to get what they wanted, when they wanted it. Corporations repealing environmental and safety laws to maximize their own profits. And against them, good and generous people who valued fairness, who sacrificed personal safety to save drowning strangers, take on Ebola in distant jungles, deliver food to starving people who shoot them for it. I wasn't naive; these same people could probably be selfish in other contexts. But they were good people.

The scientists in the Caymans probably also thought of themselves as good people, as did whatever billionaire philanthropists were funding them. They were creating and scattering seeds of heightened altruism. Enough seeds would survive to pass on that altruism, aided by the biological mechanics of a gene drive, to eventually swing humanity toward greater concern for each other, for their societies, for the future. It might be a small and scattered planting, but it *was* a planting and, in time, might spread like kudzu. The scientists were growing goodness.

Was I going to make my daughter less good because she might become too good? She might do something generous for her entire society. Or she might just become one of the everyday altruists, the volunteers at nursing homes, builders for Habitats for Humanity, neighbors you can count on to help without expecting anything in return. The true invisible, indispensable people.

After we made our decision, we went upstairs to gaze at the kids. Brady lay sprawled in his crib, one arm flung around his favorite blankie. Kenly lay straight in the bed like a miniature soldier. "This scene is such a cliché," Jen said, and gave a single sob.

After she went to bed, I stayed up, drinking a bottle of Scotch somebody gave us last Christmas, which had sat unopened in the back of the pantry. The crescent moon left the window and clouds moved in. Eventually it began to rain, a soft pattering against the pane. I opened the window to smell the spring.

2:00 a.m. That was 11:00 p.m. in California, not too late. The lead scientist, whoever she was, wanted our decision.

I picked up my cell to make the call.

**Lisa Yaszek**

NANCY KRESS (B. 1948) IS A U.S. AUTHOR OF SCIENCE FICTION AND FANTASY celebrated for her ability to tell large-scale stories about the technocultural transformation of humanity through the focusing lens of interpersonal and familial relationships. Born Nancy Anne Koningisor in Buffalo, New York, she was raised in the nearby town of East Aurora in a large Italian American family. She obtained her undergraduate degree in elementary education from State University of New York at Plattsburgh, then went on to teach fourth grade for four years. In 1973 she moved to Rochester to marry Michael Joseph Kress, with whom she had two sons, Kevin Michel Kress and Brian Stephen Kress.

Although the adolescent and twentysomething Kress did not imagine that she would someday become a writer, she was always drawn to stories of other places and times. As a child, she read nineteenth-century girls' author Louisa May Alcott and twentieth-century Western writer Zane Gray. In elementary school, she was bored by C. S. Lewis but fascinated by the collections of fairy tales that were designated as appropriate reading for girls. She discovered science fiction at the age of fifteen when she came across a copy of Arthur C. Clarke's *Childhood's End*; a few years later, the work of Ursula K. Le Guin captured her imagination as well. Even as she immersed herself in what would eventually become her chosen genre, Kress continued to engage other modes of writing and thinking. In her twenties, she grew interested in Objectivist Ayn Rand, but eventually rejected her philosophy because it celebrated an ethic of personal responsibility at the expense of community and care for others.

Kress's diverse literary interests have always been reflected in her career as an author. She began writing in the 1970s as a stay-at-home mother looking for something to occupy her time while she was pregnant with her second son. Kress sold her first story, "The Earth Dwellers," to *Galaxy* magazine in 1976. By the mid-1980s she was publishing regularly in major science fiction venues including *Omni*, the *Magazine of Fantasy and Science Fiction*, and *Asimov's Science Fiction* magazine, and she had earned her first Nebula Award for the 1985 short story, "Out of All Them Bright Stars." This period also saw the publication of Kress's first three novels: the fantasies *Prince of the Morning Bells* (1981), *The Golden Grove* (1984), and *The White Pipes* (1985). These books pay homage to Kress's interest in both fairy tales and feminist speculative fiction by casting all kinds of women—young and old, rich and poor, often with children in tow—as the "sheroes" of their own lifelong adventures.

After acquiring two more degrees (an MS in Education in 1977 and an MA in English in 1979, both from SUNY Brockport), divorcing her first husband in 1984, and starting work as a corporate copywriter and part-time English instructor, Kress published her first short story collection, *Trinity and Other Stories* (1985), and her first two science fiction novels, *An Alien Light* (1988) and *Brain Rose* (1990). These works combined Kress's already established expertise at character development with the thematic question that would guide much of her later fiction: can new technologies that transform human bodies and minds also transform human nature? In 1990 Kress became a full-time writer and began to consolidate her reputation for weaving ethical debates about the meaning and value of new scientific and medical trends into vividly dramatized stories. This is particularly evident in the Hugo and Nebula award–winning novella "Beggars in Spain" (1991), which uses genetic engineering as the occasion for two strong, stubborn, and brilliant women to test the merits of an Ayn Rand–like Objectivism against an Ursula K. Le Guin–inspired communitarianism.

In a move that would come to characterize much of her later science fiction writing, Kress developed "Beggars" into a novel and then a future-history trilogy to more carefully explore the impact of genetic and social engineering on subsequent generations of her protagonists' intertwined families. Genetic engineering and quirky characters in complex relationships are also at the heart of Kress's technothrillers, *Oaths and Miracles* (1996) and *Stinger* (1998). These stories revolve around a neurotic FBI agent whose obsession with his ex-wife compromises his career, but who redeems himself by teaming up with a series of remarkable women to solve crimes of genetic engineering that threaten the world.

The publication of the Probability sequence (1996–2002) and the Crossfire diptych (2003–2004) cemented Kress's reputation as a leading author of hard science fiction. Like much of her previous work, these series feature families whose lives are radically transformed by new technoscientific situations. In both new series, however, Kress exchanges the near-future Earth locales of her earlier future histories for the far-future alien settings characteristic of the interplanetary romance. She also makes genetic engineering and other sciences (most notably, anthropology and physics) increasingly central to her stories. This latter shift seems to have been spurred by both scientific developments including the mapping of the human genome and personal ones such as Kress's 1998 marriage to scientist and science fiction author Charles Sheffield, who died in 2002 from brain cancer.

Over the past decade and half, Kress has continued to win accolades for her science fiction stories including *Steal Across the Sky* (2009), *Before the Fall, During the Fall, and After the Fall* (2012), and the Yesterday's Kin series (2014–2018). She has also continued to publish technothrillers (including the 2008 disaster tale *Dogs*) and female-oriented fairy tales (in the form of two original stories in Anna Kashina's 2012 anthology, *Once Upon a Curse*). In this period, Kress has made a name for herself in young adult science fiction as well, with regular contributions to Dreaming Robot Press's *Young Explorers Adventure Guide* series under her own name and with the publication of the Soulvine Moore Chronicles, a fantasy series written under the penname Anna Kendall. She has published six short story collections in English and one in French; the latter, *Danses aériennes*, earned Kress a 2018 Grand Prix de l'imaginaire award.

To date, Kress has authored thirty-three books, including three books on writing. Her work has won six Nebulas, two Hugos, a Sturgeon, and the John W. Campbell Memorial Award and has been translated into two dozen languages, including Klingon. She has also been a regular contributor to *Writer's Digest* (for which she wrote the "Fiction" column for sixteen years) and a teacher for writing workshops in the United States, Germany, and Japan. She currently co-hosts Taos Toolbox, an intensive writing workshop she teaches each summer with Walter Jon Williams. Kress lives in Seattle with her husband Jack Skillingstead, whom she married in 2011, and Cosette, the world's most spoiled toy poodle.

## An Interview with Nancy Kress

**LY:** You're often celebrated for stories that link big technoscientific ideas with intimate portraits of families and relationships. Have your own family experiences factored into your writing?

**NK:** I think everything plays a part in my journey. Everything that you see or read or experience or view on television or movies, everything goes into that deep well of unconsciousness and sort of mutates down there into, hopefully, something rich and strange.

I write a lot about the relationships among sisters. I have a sister to whom I'm very close. I also have two brothers, whom I love, but I'm not as close to them. So yes, I would say my family did play into it. But it also played into it in another way. I grew up in the 1950s, and they were a much different time for women. When I was twelve, my mother sat me down and said, "Do you want to be a nurse, a secretary, or a teacher?" That was literally the entire scope of what she, with her working-class, Italian American background, could imagine for women.... So I thought it over and I said, "I'll be an elementary school teacher," and that's what I became. I think one's family background always plays into the larger life path that you take. I came to writing science fiction pretty late—[I was] almost thirty—and that was part of the reason.

**LY:** The 1950s was an exciting time in terms of scientific and technological development. Did that interest you as a young person, and did it ever even cross your mind that you might grow up to write about such issues?

**NK:** It never crossed my mind, and I was not interested. It is sort of amazing to me in retrospect. Again, this may have gone [along] with something of the science-is-for-boys-not-girls period. But I was an English major in college, and in graduate school, and it was always the narrative that interested me.... My first three novels were fantasy. And then several novels after that were science fiction, but not hard science fiction. They didn't use actual science, they only used it as metaphor and as background, in the way that a lot of science fiction does. I was, I would say, at least in my forties, before I became interested in genuine science. What caused my interest was genetic engineering, and then everything else followed from that. Physics is still something I don't feel as confident of as I do biology, even though I read about it.

**LY:** Given how central relationships and families are to your fiction, it seems significant that you tend to write about sciences that change exactly those things.

**NK:** Science gives birth to technology, and technology gives birth to societal change. And it's the societal change, especially ethical aspects of that, that interests me. The science itself is fascinating. But unless I can translate it into narrative, and its effect on people, it doesn't hold as much fascination, and it doesn't, of course, create stories. Because stories are made out of and for people.

**LY:** Who are the models for your families in science fiction? Do you draw on your own family, or other families, real or imagined?

**NK:** Very seldom my own, except possibly my sister. I guess, like everyone else who was an English major, I read all the time, and observed people, and tried to come to conclusions about them, and as I got out in the world more, was shocked by some of the families I saw. And I think all of that feeds into it.

**LY:** Do you have any favorite fictional families?

**NK:** Yes. The late Ursula K. Le Guin, who is my all-time favorite science fiction writer, said quite early on that there were not enough children in science fiction. She said this when she was writing *The Dispossessed*, which I think is one of the best science fiction novels ever written. And that struck me as really strongly true. Because much science fiction includes whizzing around the galaxy, doing whatever it is you're doing, or inventing marvelous things, but there are no children in that. It's hard to imagine anybody in William Gibson's world actually mixing orange juice for school, and packing a school lunch, and dealing with measles shots, what have you.... I didn't want to do that. I wanted—I have two children of my own, and I was a teacher of elementary school children, and I wanted to write the kind of things that focused on families, often, not always but often, with children.

**LY:** You've also talked about Arthur C. Clarke as an early influence. What interested you about his writing?

**NK:** It was the big ideas, and also the strong strain of lyrical romanticism that goes through Clarke. The first science fiction novel I ever read was *Childhood's End*. And, of course, there is a family in there. The two children... become the vanguard of the next stage of human evolution. That impressed me enormously. That book has everything in it. It has individual characters, it has families, it has the whole entire human race transforming. And then Clark destroys the whole planet at the end. Aliens—it's all there. And as an introduction to science fiction it was pretty damn amazing.

But the lyricism, the romance lyricism that goes through a lot of the short stories, like "The Star" and "The Sentinel"—there's that yearning for what's out there, at the same time understanding that what's out there may not necessarily be benign.

**LY:** A lot of your early reading was outside the realm of science fiction, as in the case of Louisa May Alcott. I would love to hear more about why you liked about Alcott and maybe who your favorite March sister was, although I bet I can guess—everyone has the same favorite.

**NK:** Jo was everybody's favorite March sister. I also liked Meg. I liked her domesticity. I have a strong streak of domesticity myself. And I liked her desire for a husband and children, which Jo never had. And I did like her. I thought Beth was sort of wimpy, and it wasn't until I was much, much older that I could appreciate Amy, and her struggle to be a great artist, and her realization that she never was going to be and having to make peace with that. But I was much, much older before I was able to appreciate Amy....

I was one of those kids who would read anything. My father had a box of Zane Grey in the attic, which he had read when he was a boy. And there were twenty-six of them. I don't know if he wrote more than that, but that's how many were in there. I read every single one of them, and it took me all the way up till about twenty-two to realize they all had the same plot. But I didn't care. They were romantic, they had the Wild West, as I'm sure it never was, they had all of these romantic plots.... I read whatever was lying around. My mother would give me books that she had had as a girl, such as Booth Tarkington's *Seventeen*, which baffled me. I didn't understand any of this guy's attraction for this baby-talking girl, no matter how pretty she was.

**LY: Have any of the authors that you're talking about influenced your own writing?**

**NK:** I don't know. It was so long ago when I was reading them. I was [under] twelve. And I don't admire Zane Grey any more. The style is overblown, the dialogue ridiculous, and the plots overly romantic. But at the time I didn't care. I read for story, I didn't read for language, I didn't read for character verisimilitude. I was interested in story. I guess that's what's lingered from that childhood: the sense of story is important.

**LY: It seems as if you're now attracted to stories that are challenging, are complicated, and make you think. That's often true of your own fiction. You don't give your characters easy outs; there are no easy heroes or villains.**

**NK:** I'm glad you said that. I almost never create a pure villain. Rarely do I create a pure villain. Once in a while. And when I'm teaching at Taos, one of the things I tell my students over and over again is that things cost. You can't just have a completely happy ending, where everybody has sailed through and come out the other end and it hasn't cost them anything. Things cost, and you have to show that ambiguity of real life in your fiction.

**LY: What about the fantasy that you were reading?**

**NK:** I read the first Narnia book, and it was okay, but I was young, and it didn't strike me with an intensity. I read all the fairy tales. Those struck me more. I remember reading Hans Christian Anderson, who horrified me. Those are horrifying.... But I read them over and over. The little girl who sinks through the swamp to the Marsh King and is brutalized down there, a lot of it was... "The Little Match Girl." ... A lot of it was horrifying. But at the same time, I read them over and over again. There must have been something in there I wanted. I don't know what.

**LY: At different points in your writing career you've returned to fantasy and fairy tales. What is it that interests you in those kinds of stories?**

**NK:** They work on a metaphorical level. My rewriting of "Sleeping Beauty" has to do with everybody else falling asleep except beauty. And there she is one hundred years, because she doesn't age.... Well, she does age, but she doesn't die. But she's the only one awake. And how do you deal with being the only person in the world? And then how do you deal with being old when everybody else awakes around you and is still young? I was interested in using the fairy tale form to explore real questions in our world in a metaphorical kind of way. All of those rewritten fairy tales, Ellen Datlow was doing a whole series of books of them, and she kept asking me for one. So, I would pick one and try to find an angle on it that everybody else hadn't already used. And that's how those stories evolved.

**LY: In addition to science fiction and fantasy, you write technothrillers and young adult science fiction. Do you balance the big ideas and the intimate relations similarly in these modes of storytelling?**

**NK:** I don't think I have enough flexibility to change my balance with the subgenre. I always try to balance the character stories with the science stories. The one I'm writing now has very complicated science having to do with physics and I have to get a lot of the science in there or the novel won't make any sense, but I can't just bore the reader with page after page after page of scientific explanation. So, you pull out every trick you know to balance the dramatic stories of the characters with the science that has to be in there for the rest of the novel to make sense. In some places this comes easy, and in some stories it comes hard.

    That's the basic problem of hard science fiction. How do you get the science in there in a way that is convincing? And to be convincing, it has to be pretty detailed. That is, both be convincing and yet balance with the story. That's what I've spent my entire career trying to learn to do.

**LY: Does organizing your stories around families and other intimate relationships allow you to do something with science and technology that you might not be able to do otherwise?**

**NK:** I think parents have an intense investment in their children, and partners—romantic partners, sexual partners, families—have intense investment in each other. It allows me to give an intensity to the ethical aspects of science fiction more than the adventure ones. I prefer that. Ursula K. Le Guin said, "Sometimes the more there is going on in the outer space, the less there is going on in the inner space." With families there's a lot going on in the inner space, because there's all of these intense personal relationships. And that's a good arena for exploring the ethics of sacrifice, and competing needs, and things like that.

**LY: Your characters are sexual beings as well as scientific and social and familial beings. What I find particularly striking is that the characters who think and talk the most about sex in your stories are not perfect Hollywood ideals—instead, you've got everything from angry teenagers to globetrotting grandmothers. In your worlds, sex is for everyone! What led you to that choice?**

**NK:** Because sex is for everyone. And because it's an important part of life, especially if you're an angry teenager. It occupies a lot of your thoughts.

    It seems to me that if you're going to create realistic people, both the sexual feelings and the feelings of love that you have for partners [need to be included]. . . . Humans have a desire for connection, I think, and the connection can be romantic, or sexual, or both, or fraternal, or sororal,

31

or familial, and I write about the need for connection that people have. I also think that a majority of human beings, if they can't get a positive connection, would prefer a negative one to no connection at all. So, there are negative connections, too, in my fiction.

**LY:** Throughout your career, you've been interested in human evolution and the ways families do and don't change over time in relation to scientific and social events. Are there any specific developments you see on the horizon that you're either really apprehensive or really excited about right now?

**NK:** The thing that makes me the most apprehensive right now... well, two things. Climate change, of course, and the other one is income disparity. As long ago as 1992, when I wrote "Beggars in Spain," I was dealing with the question: as automation and computers and software and nerds run more and more of society, what happens to all of the people at the bottom who used to do those jobs? In 1992 it was a concern. Now it's getting to be a major, major concern.... The truth is, there's about a third of the people in this country that we don't actually need economically. But they're human beings, and what's going to happen? How are we going to restructure society to cope with this? Because so far, we're doing just as bad a job of that as we are of dealing with climate change.

What I'm excited about is genetically altered crops. We've made enormous strides on this, on GMOs [genetically modified organisms], and we are making even more, and we're going to need them. If the planet is really going to swell to nine billion people, and we're really going to have climates change through desertification or rising seas or growing warmth where there wasn't that much warmth before, we're going to need to engineer crops that can feed all of these people and that can adapt to new environments quicker than natural evolution would have them adapt. And GMOs are doing that. They're developing tomatoes that can grow in more brackish and saline water than they do now. They're developing all kinds of crops that will be able to feed a burgeoning population under rapidly changing soil conditions and weather conditions. I'm very excited about this, and I don't understand the anti-GMO people.... I have a novella coming out, sometime next year probably, a stand-alone novella, and it deals exactly with this, with GMOs. And it's going to be controversial because GMOs are controversial.

**LY:** One last question. *Who* interests you in science fiction right now?

**NK:** There are a number of the young writers coming up that are very good. I like Sarah Pinsker. I think she's done some really interesting things. I like Matthew Kressel. I like Rebecca Roanhorse. And I'm especially excited about Suzanne Palmer. Her novelette, "The Secret Life of Bots," managed to do something new with robots, and something charming with robots. In fact, when I teach at Taos this year, I'm going to use it because it is an almost perfect example of raising the stakes, scene by scene, until you get to the end. It's also an example of making a character charming and interesting without making a robot too humanoid. And that's not easy to do.

**LY:** Do these authors have anything in common beyond their interest in science fiction?

**NK:** They all have a sense of humanity, even the ones writing about robots. Characters are, for me, what drives fiction, even more than plot. If I find a character interesting, I will read an entire novel in which almost nothing happens.... So, all of those writers create characters that grab my attention.

# 3 ECHO THE ECHO

**Rich Larson**

"YOU'RE NOT WEARING THE WEB," I SAY, WHILE MY GRANDMA LAYS OUT SLICES OF chocolate chip banana bread and bright orange cheddar cheese on the table.

"No." She casts a dark look toward the sunken couch, where the neural web is sitting on top of its packaging like a desiccated jellyfish. "Ben would spit on me from heaven."

It does look a little out of place in her apartment, which has gone more or less unchanged for the past two decades: thick beige carpeting, stucco walls, an ancient cuckoo clock, and a wicker basket of sun-curled *National Geographic* magazines. Hardcopy photos of the family hang on the walls, and her other decor is comprised mostly of disturbing porcelain babies.

Technology's creep is inevitable, though. Some of her ferns are in smart pots, scuttling around the window to follow the slanting sunlight, and behind us there's a screen she keeps defaulted to a world map. She uses it to keep track of us. I can see my sister's route from Newark up here to Ottawa, my own zig-zags through Colombia and Bolivia last summer, and my usual home base out in Edmonton.

"You have an umbrella," I point out. "Mom bought you a nice umbrella last year."

"Open this," she says, handing me an unopened 2-liter bottle of ginger ale. "Oh, your mom. She's always buying me things, isn't she? Always knows best for the whole gang."

I grip the cap and twist, and the carbonation rumbles up under my hand. She's been buying the same big green bottle of ginger ale since I was six and walked to her house every week in the summer to eat banana bread and play Scrabble. I don't know where she finds it now—plastic bottles are mostly extinct.

"You know what I like to do," she says as I pour two glasses, "is to add—"

"Some cranberry juice," I say. "I know. You sit. I'll grab it."

She sets down in the chair like a bird perching in high wind, making little adjustments, false starts. She's clumsy lately and hates it, so I pretend to be searching hard for the cranberry juice. When I come back she's slumped back holding her bad wrist. Her good one has her phone dangling from a rubber strap.

"You know, I can barely feel with this hand anymore," she says. "Barely feel a thing."

"So," I say, pouring a splash of red into her glass. "Why would grandpa spit on you for wearing the web?"

"It's not the Christian thing to do," she says. "Leaving the echo behind. I've decided. It just isn't right. The Lord gives us our time, and that's that. I'm ready to go."

"You just don't like how it looks."

"Oh, I hate how it looks." She shoves the banana bread over to me. "It's like wearing a dunce cap. But that's not the important thing. The important thing is how it affects the soul."

I take a slice of banana bread and reverse-sandwich it with cheddar cheese. "You've been streaming sermons from those Korean megachurch assholes."

"Don't use that language."

I take a bite. "How's it affect the soul?"

"Nobody knows," she says, triumphant. "Nobody knows, and that's why it's such a bad idea."

"You're worried it's sucking your soul right out of your scalp, huh?"

"Maybe."

I set the banana bread down so I can put my hands on my head. I spray some crumbs when I make the vacuum noise. "Come on," I say. "You were a nurse. It's medical technology."

"I don't like all this new technology."

I point at her phone. "You like WordWhirl."

"You haven't started a game with me in years. We used to have five, six games going at a time." She pauses, conspiratorial. "You know, I beat your mom once. Leonine. That was the word that won it. Lionlike. She's so competitive, your mom. Always has to win. Always has to be right."

"Yeah, she's the worst." I whisper it into my ginger ale, like Mom might be listening in, and that gets her to chuckle. "The web's not taking anything from you. It's just copying. It's copying your brain. It'll be like a photo album, but better."

"It'll talk," she says, sounding vaguely repulsed.

"You love talking."

"I do not. You know who loves talking?"

"Tina Reichert."

"Tina Reichert," she agrees. "Invite her for tea at ten o'clock, she stays until supper."

"Well, the echo won't be like Tina Reichert," I say. "We'll be able to turn it off."

"That sounds even worse," she says. "Like I'm just a program." She rubs her bad wrist. "This hand, you know, I can't feel a thing with it. Not a thing."

"Yeah."

We sit without talking for a little while. I look over her wispy white head at the map, where she once traced her childhood for me, fleeing from Ukraine to Germany to England to Canada during World War II. Maybe she's thinking about all the dead people she knows who never had the chance to get an echo: her older sister Maria who starved during the artificial famine, her older brother Fritz who had a bone disease, the boy she loved who got shot off a motorcycle. Those are the ones I remember, but I know there were more.

"It's for your great-grandkids, you know?" I say. "And the great-great-grandkids, et cetera. They'll get to see you on Christmas and Easter and their birthdays and stuff. They'll get to hear your stories."

"But it won't be me," she says. "I'm going to be like Tupac."

"You're going to be like Tupac?"

"A hologram."

"Oh. Yeah. Yeah, it'll be a hologram."

She stares balefully down at the banana bread. "I know your mom sent you. To persuade me. Here, eat some more."

I take another slice. "Is it working?" I ask. "Should I sing and dance?"

"I've been thinking about it a lot," she says. "During the night, especially. You know my sleeping pills ran out? So I lie there awake, thinking." She looks up at me, and I feel a jolt of fear in my stomach because her eyes are wet. She doesn't cry. Not since grandpa died, at least. "I'm already an echo," she says.

I put my hand tentatively on her good wrist. "What do you mean?"

"I did wear the thing for a while," she says, nodding toward the cast-off web. "What it does, it records everything I say, everything I think. All day. And then I can review it."

My stomach goes submarine. "Oh."

"I've been reading the same damn poem over and over," she says. "In that book of German poetry your mom bought me. I keep picking it up and reading it."

"You like it."

"I keep saying the same things over and over. It's awful. I don't realize it. I'm probably doing it right now. I keep telling the same stories and whining about the same aches and pains."

"You're ninety-seven," I say. "You're allowed to whine all you want."

"At first I thought I must have dementia," she mutters. "But apparently it's normal, at my age. The forgetting and the repeating. I'm just old. I'm not who I used to be."

"You're plenty."

She glares at me. "And if it's not Christmas, if it's not someone's birthday, you think your mom comes to visit me? You think anyone comes to visit me? Unless it's an occasion, I'm just sitting here alone, thinking the same things over and over. Like a record—you know records? How they get worn out from playing them too much? That's me. So the echo would be an echo of an echo."

I don't know what to say. She got me. Here's the truth: Since I moved out west I only come back to Ottawa for a few weeks every year to see family and old friends, usually combining it with a trip to my sister's place in Newark. I know I used to face-call grandma every week, but thinking about it now I can't remember when the last one was.

I have conversations *about* her, usually about her health, about hip replacements and UTIs and sciatica, but those are with Mom, and sometimes it's like we're talking about maintaining a car. And of course it was Mom who put me up to the job of convincing grandma to do the echo program.

"I'm sorry," she says, and I hate that because I should be the sorry one. "Let's talk about something else. How long are you here for? Until Tuesday, your mom said?"

"Yeah," I say. "Going to visit friends in Montreal for a couple days, then fly back to Edmonton on the Friday."

She pushes the ginger ale across the table. "Pour yourself some more. Better you do it. I'm so clumsy these days." She leans back, rubbing her bad wrist. "You know, the fingers on this hand barely feel a thing anymore. Not a thing."

"Yeah," I say. "I know."

I leave my grandma's place with a chunk of banana bread wrapped in wax paper, promising to face-call her more often. Outside it's gotten dark, and it's raining the big slippery sheets that turn the icy sidewalks into a death-trap. I should have asked to borrow her umbrella. Springtime in Ottawa: an endless freeze and thaw, skin-prickling humidity, late snowstorms that get worse every year. Even ninety-seven-year-olds believe in climate change now.

I had vague plans to meet friends downtown later, but when my phone buzzes I pull it out to find a message from my avatar: *Hey, brother. Got you a date.*

I open the dating notification to see the profile. A fat raindrop splatters the screen and obscures her face, and that kind of date seems perfect right now. With friends I would feel the need to work through this whole echo conversation I just had with the woman who halfway-raised me as a kid. With a stranger I can just not think about it.

I duck into a bus shelter and take a better look at the situation. Her name is Ana and our avatars have been hard at work ever since I got into town a few days ago, evaluating compatibility and churning out a dummy conversation. I skim through it. We've been talking about Colombia, where her dad is from and where I vacationed last summer, and complaining about the weather in Ottawa. But also talking dirty in Spanish, which is an interesting development I've never seen before.

She's photo beautiful—apparently she does some modeling as well as working for the Canada Revenue Agency, so her main photo is from a shoot in a warehouse somewhere, backlit eerie Matrix green. Her textured black dress clings hard to the curve of her back, and she's doing that wrist-on-the-forehead pose models sometimes do. Her hair is dyed red, and her lipstick is the same shade.

There's a loop of her at New Year's opening a champagne bottle surrounded by people facially similar to her, probably family members, and another of her posing with a purple beach ball by a sunny pool at a Vegas hotel. There's a tattoo on her tanned ribcage, letters I can't make out.

She lives across the border, all the way out in fucking Aylmer, Quebec, but suddenly I've always wanted to go to Aylmer, Quebec. Our avatars have it all worked out. At 8:30 p.m. we're meeting at Bistro Mexicana 129, because there's no Colombian restaurant, and from there the date can branch to a British pub around the corner or to her place.

I hope her place, partly because driverless cars are still illegal in Quebec and it can be hard to get a traditional Lyft all the way back into Ottawa in the middle of the night. My phone buzzes again, my avatar prompting me to order my first ride. It takes a half hour to get to Aylmer with standard traffic, and it's 7:58 p.m.

I'll see friends tomorrow. I order the ride. A pictogram of the car drifts and swivels on my phone's map, heading my way, and I put the whole echo thing as far out of my mind as possible.

BISTRO MEXICANA 129 IS EMPTY ASIDE FROM AN ELDERLY COUPLE IN THE CORNER and a black-haired waitress. I try her in Spanish and get a blank look, then switch to English even though I really should be talking French since I'm out here in Quebec. I tell her it's my first time in Aylmer and she says in summer it's nice and we make some small talk while I select a table for two just off the bar.

Ana shows up in her work clothes, some kind of beige pantsuit thing, government ID on a lanyard around her neck. "Sorry I'm late," she says. "Had to steal my stepdad's car to get here. Not his good car. The shitty one."

I'm confused until I remember the no self-driving cars in Quebec thing. I get up for a hug and then we both sit down. She's not as beautiful without angles and airbrushing, but who is, and in real life she smells good and has this sexy kind of smirk like she's remembering a dirty joke.

"Why didn't you steal the good one?" I ask.

"He still hasn't forgiven me from last time," she says. "When I was seventeen I took his Mercedes to a house party and cops showed up and we left in such a hurry I swiped all the paint off a parked car. Like this big huge stripe. I tried to tell him it could be a racing stripe." She picks up a menu. "I'm normally a lot prettier for dates," she says. "Worked late today. Budget stuff."

"I'm normally way prettier too," I say, taking my toque off. "Usually I spend two hours minimum on my hair. Today I did only one."

She does the smirk again. "Only one, huh. Slacker."

"One and a half."

"You're pretty pretty anyway," she says, and from the way she says it it becomes clear we're going to fuck later.

She orders a Purple Rain cocktail, which looks like a fishbowl full of purple Kool-Aid and ice cubes. I get some kind of Quebecois IPA. We clink drinks.

"You have to drink it fast," I say. "Before the fish dies of alcohol poisoning."

"There's no alcohol in this," she says. "Just codeine."

I can see why our avatars jibed. We have the same sense of humor and the same kind of stories from our partying days, though hers blow mine out of the water: accidentally taking a limousine from Las Vegas to Salt Lake City with a trio of high-class hookers, forgetting she had drunkenly bought a baggie of coke in Honduras and only finding it in her bag after she got through airport security in Montreal, and once getting abducted by aliens.

"Either that or I broke into the Biodome," she says. "But I disappeared from the club and showed up half an hour later at my friend's place covered in burrs."

"In birds?"

"Burrs. And I had twigs and leaves and stuff in my hair, and I kept saying I ran through the forest. But there's no parks around there, nothing except the Biodome, and the Biodome has a huge fence." She shrugs. "I do like to climb stuff when I drink."

"I'll find us a park."

She does a snorting kind of laugh, then takes out her phone and thumbs furiously for a few seconds before she looks up. *"Hablas espanol, verdad?"*

*"Hombre, claro."*

*"Que hiciste en Colombia? Fuiste de vacaciones o que?"*

*"Si, vacaciones,"* I say, doing an exaggerated sniff. "You're only just reading the dummy convo now? You must really trust your avatar."

"I mean, that's the whole point," she says. "They know us better than we do, right?" She drains the last of the purple concoction and wipes her mouth. "I think I want to get you drunk."

I point to the empty bowl. "You swallowed the fish."

"Fuck yeah, I did."

She asks the waitress for tequila, any kind of tequila, and when the waitress asks our price range I say, because I suspect I'm paying, please think of us as poor.

"So you live with your stepdad?" I ask, because it's time to start working out logistics.

"No," she says. "I live at my place. With my four-year-old, but he's at grandma's for the weekend."

"Fun," I say. Usually my avatar would notify me about someone having a kid, but my avatar also knows I don't live here. So does hers. They know it's just tonight.

"Yeah," she says. "I've got phantom four-year-old syndrome. I feel him walking along behind me or grabbing my hand, then I turn around and he's not there. It's freaky."

The tequila shots arrive on a little wooden board with salt and slices of lime. We clink them together. It goes down smoother than anticipated, and then we're off to her place.

ANA CLAIMS HER APARTMENT IS A DISASTER ZONE AND SHE CANNOT, IN GOOD conscience, let me inside until she cleans up a little.

"So here's the plan," she says. "You go grab us a bottle of wine. At that liquor store." She points out the glowing sign through the freezing drizzle. "It takes ten minutes to walk there with a four-year-old, probably half that without one. So I'll have time to clean up. Oh, you might meet Patricia!"

At this point I'm buzzed enough to accept all sorts of plans. "Sure. Who's Patricia?"

"She works there," Ana says. "She's my best friend."

"You smell good," I say.

"Thank you," she says. "I take no chances with that shit. I've got no sense of smell, so I'm always wearing perfume."

"My grandma lost hers, too. She made me swear to always tell her if she stinks."

She taps the tip of her nose. "It's 'cause I busted my nose."

"Too much coke, huh. Ruptured septum."

"Not even. When I was five I broke my face running into a wall. I thought it would open up and take me to Narnia." She traces the bridge of her nose. "So this is a little crooked and I can't smell stuff and there's a little dent in my forehead. You can touch it if you want."

I reach over and feel the small divot above her eyebrow.

"I'm really good with my angles," she says. "So the crookedness doesn't show up in photos."

Then we clamber out of her stepdad's Toyota Corolla and part ways: she darts through the puddles to the apartment door; I flip up the hood of my coat and start trudging down the road. I'm insulated by the alcohol and by the fact that I'm kind of fascinated and infatuated with this Ana who got abducted from a nightclub by aliens and tried to run through a wall to Narnia. Avatars always know, somehow.

A delivery drone stutters through the rain up above me, illuminated in orange flashes as it passes the streetlamps. The space and emptiness remind me of Alberta—infrastructure designed for cars, not for people.

But the walk goes fast without a four-year-old, and few minutes later I hop across the four-lane, cross the parking lot, and walk into the liquor store. A camera scans my face for ID on my way to the cooler.

The cashier is a woman with a red shirt and salt-and-pepper hair in a raggedly fashionable cut. "Hey," she says when I put the Riesling down. "How's it going?"

"Good," I say, checking her nametag.

"You look sad."

That catches me a little off-guard. "I always look sad," I say. "Bad face muscles, I think. You're Ana's friend."

"Huh?"

"You're Patricia." My phone blips the payment over. "You're friends with Ana."

"Oh. With Ana Marie. Yeah, I am." She gives me a more scrutinizing look. "You dating her?"

I do a drunk kind of smirk while I put the bottle in my coat. I feel the wrapped up slice of banana bread that I'd forgotten about. "Just friends."

"You'll need two bottles for Ana," Patricia says. "I recommend two bottles."

"It's okay," I say. "I also have banana bread."

I CALL ANA FROM THE STOOP AND TELL HER I'M FROM THE HURRICANE INSPECTION committee, heard about a real bad one inside her apartment, and she buzzes the door open. I jog up the stairwell, which has steps capped in brown rubber and a strong smell of weed. She lives in 301; when she opens the door, she is wearing pink pajamas instead of her work clothes.

"I met Patricia," I tell her. "She said I look sad."

Ana grabs the wine out of my unzipped coat. "She says that to everyone."

The apartment is still a disaster zone, mostly because kid's toys are scattered all over the place: books and blocks and action figures and a little holotable I would have loved to have when I was little. One of those annoying yellow balls that jumps and rolls on its own is meandering around the floor. Clothes are heaped in the corners. A Colombian flag covers one window.

I trail Ana to the kitchen, where she's retrieving wine glasses. There's a designer purse sitting on top of the stove; the stove clock is about three hours behind. Several big flower bouquets are on the counter beside it with the delivery tags still on. Marilyn Monroe is all over the walls in black and white, which is never a good sign, and empty wine bottles abound.

"I never bring people back here," she says. "Definitely not on the first date."

And I think, *she says that to everyone*, but I don't say it because I know a lot of people still like to play pretend this way, and who knows, she might even be telling the truth. She does have a kid.

"I don't even know what to do now," she says.

I kiss her, and things continue from there. We end up on the couch playing truth-or-dare with ancient reruns of *Bob's Burgers* playing on the screen in the background, no doubt recreating some high school hookup of hers, and the wine goes fast. She pulls some beers out of the fridge. We have a businesslike discussion about the pros and cons of bed versus couch with my hand under her shirt, then migrate to her bed.

The shirt comes off and I see that tattoo from the poolside loop, characters I don't recognize underneath her left breast. "What is that?" I ask.

"My son's birthday," she says. "In Arabic. My first boyfriend was Arabic, and he was an asshole. He's in jail. Then I slept with his friend. Then I dated an Italian guy for a while. You're number four." She peels my shirt off and gains a look of mild disappointment. "You don't have any tattoos."

"Never had anything important enough," I say. "I always think, maybe when someone really important to me dies."

She makes a face.

"I get your thing, though," I say, sliding off my pants. "I get loving your kid. That's like a biological compulsion. People love their kids like crazy."

"Shut up," she says. "Don't talk about kids while we're having sex."

And I get what I want, which is not the sex but the forgetting, the moment where you can't think about anything, not looking sad or people dying or getting old or any of it. It's drunk and sloppy but when we work together on her clit she manages to come a couple times too, and then we lie there together for a long time. Eventually she peels herself off me and heads for the bathroom. She comes back setting an alarm on her phone.

"I gotta get up super early to return my stepdad's car," she says. "I carpool to work with him and my brother. You can keep sleeping."

She clambers back into the bed and wraps herself around me. There's still that tender feeling, the hormone serotonin afterglow that fools you into thinking you know a person. So as I'm drifting off I ask her if she knows about echoes.

Her hand stops stroking my shoulder. "What?"

"This brain-copying technology," I say. "My mom's a neural engineer. You know how our avatars work? Like, reading all our messages, all our ad info, then saying whatever we would say? This is the deep version of that. The raw material isn't text. It's actual brain activity. So the echo thinks how the person thinks. They're doing a pilot program now for the compact model. A take-home thing."

She rolls over. "I know about echoes. And the pilot program. Go to sleep."

I try, but she snores like a motherfucker.

IT'S 11:07 A.M. WHEN I WAKE UP FOR REAL, AND ANA IS LONG GONE. I CAN FEEL the start of a cold sitting in the back of my throat, and I'm pretty sure she gave it to me. Between her snoring and her hogging the blanket, I didn't sleep so well—to be honest, I never sleep so well with someone else in the bed—and I was still half-awake when she left earlier in the morning. I dimly remember a conversation where she griped about not having time to wash her hair and asked for input on her outfit, wondering if she could get away with shredded jeans because it's Friday.

I badly need to shit, so I roll off the mattress and head straight for the bathroom without bothering to find my strewn-everywhere clothes. Her bathroom did not escape the hurricane: every single surface, from the counter to the top of the toilet to the edge of the tub, is crowded with eco-friendly cleaning products, candles, cologne bottles, lotions, razors, scissors, pads, a kid's Dalmation-shaped toothbrush, a curling iron, makeup, and deodorant bars. And yet there is no toilet paper on the roll, which of course I notice only after I'm done shitting.

"Goddamnit, Ana," I say, then start hunting. There's none in the cupboard under the sink and none in the squeaky-hinged closet. I'm checking the plastic drawers for wet wipes when a familiar size of baggie catches my eye. I pick it up, try to guess how many grams it is. The smell of cocaine always reminds me of something my grandma used to bake peppermint cookies with, not so much in the smell itself but in the way it stings your nose.

I absentmindedly crush the lumps out of it, then since I'm still a little drunk I open the baggie and take a bump off my thumb. It occurs to me that the plastic drawers are definitely within reach of a four-year-old. I go grab my phone and message Ana to ask if she minds me taking a bump.

*The four grams in the bathroom? Don't, I'm selling it. I can get you some though.*

I ask her about the toilet paper.

*Check the pantry maybe. I forgot my gov ID; can you bring it to my work? Haha.*

I take the coke with me into the kitchen, still buck naked with an unclean asshole, and find a spot for it up in a high cupboard beside a wooden pepper grinder. Then I open the pantry. The first thing I see is a box of animal crackers, which makes my stomach give a nostalgic churn of hunger. More importantly, there's an open package of toilet paper. I grab a fluffy new roll and am about to head back to the bathroom when something else catches my eye. Up on the highest shelf: a slice of familiar packaging, a tangle of dangling electrodes.

I remember what Ana said when I asked her about echoes, and what she said about her kid being at grandma's for the weekend, and I figure her mom must be in the pilot program, must be dying of cancer or something. Maybe she's trying to convince her to wear the web the same way I'm trying to convince my grandma.

The coke has me feeling jittery and mischievous so I reach up and grab the hardware. I'm curious about the playback, about what my grandma saw that made her so sad. The web is too small to fit over my head, but there's an option for holo projection so I do that instead.

The holo flickers to life inside the pantry; I shut the door behind me to see it better. It's showing Ana in the kitchen, but from a weird low angle, like someone sitting on the floor.

"So first we're going to your appointment," Ana is saying. "Then we're taking you to grandma's house. And you'll be there for two sleeps. Did you pick the toys you want to come along?"

"This one." A four-year-old's hand holds up a plastic dinosaur, and I realize whose perspective it is and why the web was so small. "I love it."

"Good choice, honey," Ana says. "I love that one too."

When she reaches down to ruffle her son's hair, her eyes are swollen and pink and she's trying really really hard to smile.

I TAKE THE OTTAWA VIA PORTAGE BUS BACK ACROSS THE BORDER. I'VE GOT ANA'S work ID in my hands and I keep looking down at it, at the solemn black-and-white photo of her that was taken straight on and does not hide her crooked nose. Once I put the child-sized web back where I found it, it wasn't hard to figure out the rest. I saw the bouquets on the counter again, the pamphlets for some support group, and a bunch of scrawled notes on a pad of paper from a conversation with a specialist that made it clear the chances of her kid making it to six years old are vanishingly small.

A message from my avatar pops up. *Hey, how was it? She's fun, right?*

I swipe it away and check the address Ana gave me, find the right stop on the bus map. I can't help but think that it knew, somehow, about her kid's echo and that's why it sent me on a date with her. It makes me want to smash my phone against the seat in front of me. The hangover is coming now, drumming into my skull.

Ana messages again. *Want to see a movie this eve? I think we talked about doing that and I'm free!*

I start composing my blow-off, something about how fun she is but how I'm in town for only a couple more days and have family stuff to do. I can't see her again, because I'd spend the whole time feeling sorry for her, imagining her five years from now talking to the echo of her dead kid about his favorite dinosaurs. The reason I spent last night with her was to forget things; the reason she spent it with me was to forget way worse ones.

I delete the message. I don't know what to tell her. I don't know anything about anything. My thumb works through my contacts and finds my grandma. It hovers for a second. Two seconds. Pushes.

"Hello?"

"Hey, grandma. It's me." I turn the video on and her deep-lined face shows up on my screen. "How are you?" I ask. "How's your wrist?"

"Oh, you know. Some days are better, some days are worse." She squints. "Where are you? Are you on a bus?"

"Yeah," I say. "I don't remember grandpa."

"What?"

"I don't remember him," I say. "I was little when he died. I pretend I remember him. I like the photo of me and him playing Legos. But I don't remember him."

"Sure you do. He made the swing set for you and your sister. You spent all day with him while he did it. You handed him tools."

"I know," I say. "I just don't remember. I still love him. But I wish I remembered him." I swallow hard. "When I visited here last year, at the end of it Mom drove me to Montreal. We stopped at the cemetery. We saw his tombstone. And the spot beside it. Your spot. And I just started crying."

"Oh," she says, faintly, looking over my shoulder. "Oh, my darling boy."

"And when I looked over, so was Mom." I wish I hadn't put it on video. I can feel the tears creeping up my ducts again. "Just from thinking about it. From thinking about not having you around anymore. I know you feel like you're not yourself anymore. Like you're already an echo."

She gives a wet laugh. "Did I say that?"

"But we're all echoes," I say, thinking about all the people my avatar and Ana's avatar churned through to put us together for one night. "We only exist when other people let us. When they need us for something. And we all have things we do over and over. Some good things, some bad things. So we're all echoes."

"Is that supposed to make me feel better?" she asks.

"I don't think anything makes it better," I say, picturing Ana's puffy pink eyes in the holo. "Not really. But the reason Mom wants you to wear the web, the reason we all do, is because we're not ready for you to die. We're not ever going to be ready. We'll look at photos and old letters and we'll tell stories, but it won't be enough. We want to hold on to as much of you as we can. The echo won't be enough either. But we want it anyways. Even just an echo of an echo."

She's quiet for a long time. "Maybe it just makes it harder," she finally says. "Maybe it's better to forget."

"Maybe," I say. "I don't know. But people aren't that rational. We want to hold on. So please, just wear the web."

She squeezes her eyes shut. Her liver-spotted skin looks as thin as tracing paper. "I love you, you know. I'm ready to go, but I don't want to leave you all."

"I love you, too."

"I'll wear it," she says. "I'll wear the dunce cap. We're having supper at your mom's tonight, I think."

"Yeah," I say. "Yeah. I'll see you there."

THE BUS DROPS ME ON BANK STREET AND I WALK TO THE CRA BUILDING FROM THERE. It looks as if they share the high-rise with a few other government offices: the lobby is big and shiny and has a lot of signage, along with a surprising number of people just milling around. I send Ana a message, then go and sit on one of those weird backless couches.

I watch workers going up and down a double escalator and start wondering about their private lives, wondering how many of them have secret miseries like Ana or not-so-secret ones, how many of them care about someone who's not long for this world. It's overwhelming, that feeling you get from realizing everybody has their own little universe of hurt and happiness. I understand why we keep people compartmentalized and let our avatars sift through them and talk to them for us.

Before I can get too existential, Ana shows up on the escalator. I watch her walk through the scanner and say something to the security guard, who laughs. In heels she's as tall as me and her hair is immaculate; she moves precisely, her chin held high as if nothing could possibly bother her. It's hard to imagine her in her hurricane-messy apartment. It's hard to imagine her walking with her four-year-old to the liquor store and trying to cope however she can with the fact that he's maybe turning five, never turning six.

"Hey, handsome," she says. "You got it?"

I hand it over stealthily, pretending it's a drug deal.

"Thanks," she says, slipping the lanyard over her head. "So stupid we have to wear these. They scan our faces anyways. Government, right?"

"Government," I echo, and for a second I want to tell her what I found in the pantry, but I don't. "I'm only in town for a couple more days," I say instead. "Don't think we're going to catch that movie."

She nods, blank-faced. "Yeah, my avatar said the same thing. One night only. It was fun."

"Let me know if you're ever out west." I take the squashed banana bread out of my coat pocket and hand it to her. "My grandma baked this. It's really good."

She raises an eyebrow, but also does that smirk. "Okay. Weird. But thanks."

We hug, and she smells good, and I figure I'll never see her again. But that's life, and in a way I've already got her echo and she's got mine. That's better than nothing.

# 4 SPARKLYBITS

**Nick Wolven**

THE CONTRACTUALLY MANDATED MONTHLY MEETING WAS NEVER EXACTLY A RELAXED affair, but this one, Jo decided, promised to reach superlative levels of awfulness. All morning she'd been marching upstairs to Charlie's room, standing at the door with her finger on the pingpanel, then chickening out and clumping back to the other moms. Failure. That was the message they sent with their silence, staring up from the table with their diamond-hard eyes. The ultimate modern middle-class hazard: a public exhibition of parental failure.

"You've gotta do it, Jo," Aya said after the third attempt, sipping the latte she'd been nursing all morning. "You just have to prepare him."

While Teri, at the ovenex, turned, nails glinting, and said in a voice that *sounded* concerned without actually *being* concerned, "It'll be easier on him, in the end."

Jo checked what was left of the brunch. No pastries, no cinnamon buns, no chocolate in sight. Just a few shreds of glutinous bagel and a quivering heap of eggs. They usually did these meetings at Reggio's, and Reggio's, say what you will about the coffee, was a full-auto brunch spot with drone table service and on-demand ordering and seat-by-seat checkout. Which was all but vital when the moms got together, when the last thing you wanted to worry about was who got the muffin and who bought organic and who couldn't eat additives or sugar or meat. Whereas when they did these things at the house, the meal always became a test of Jo's home-programming skills. Likewise the coffee prep, likewise the seating, likewise every other thing.

All she needed, Jo thought, was one tiny bite of cinnamon bun to help her through. But a rind of hard bagel would have to do. Wedging herself into the chair by the dynawindow, Jo blinked away the backyard view and called up the scheduler.

Ten-thirty.

Half an hour to go.

Sun Min came to the table, blowing holes in the foam on her third cappuccino. "Sooner you get started, easier it'll be. All the forums recommend the same thing. Groundwork."

Jo gnawed off a piece of bagel. "I'm just not sure this is the best thing for him."

The atmosphere tightened. Looking around the kitchen suddenly felt like staring into a stranger's frozen smile. Today was more than a family meeting, Jo reminded herself. More than a test of her mommying skills, more than a battle for Charlie's future. It was a performance.

Naturally, it was Teesha who spoke next. Gentle Teesha. Supportive Teesha. Political-candidate-advising Teesha. If Aya was the perfectionist of the group, always playing CEO; and Teri the consummate TV exec, with her pore-rejuve treatments and her million-dollar hair; and Sun Min their literary sophisticate, then Teesha was... well, what could you say about Teesha? She had the skills to handle lobbyists and congressional candidates, the biggest tantrum-throwers in town. Surely she could have handled kids, too, if she'd felt like blocking out the time. Instead, she'd taken it on herself to handle Jo, playing the helpful grandmother to Jo's perennially flustered mommy. Which was nice, to be sure, and not to be scorned. But Jo could never figure out the best way to react to all this meta-mothering.

Teesha took Jo's hand. In this family of megamoms, this clique of accomplishers—the Queen Bee, the Glamatron, the Editrix, the Matriarch—Jo was just Jo, the standard model. The person who pulled slop out of the ovenex, switched on the evening TV stream, and took a slow dive face-first into the loveseat. Not much to say about good old Jo, except that she was here. Always here.

"You scared?" Teesha used a voice that had probably helped trailing candidates through poll-number crashes, brought congressional aides down from caffeine-pill binges, coaxed suicidal interns off hotel ledges. "Worried you might blow it? Jo, let me tell you something. We're *all* scared. We've seen the therapist reports. We're watching his numbers. Yeah, we're worried about Charlie's progress. But that's why we're here. To tackle this together."

"But you—" Jo checked herself. "You don't have the relationship with him I do." Big mistake, she thought, looking at their faces. "What I'm trying to say is, it feels like *I'm* always the one who has to—"

Teesha eased back, using little plucks of her fingers to resettle her Yoruba-patterned shawl. A trick of hers, the gesture was saturated with authority.

"We're a family. That's what we agreed. Equal partners. The contract says—" Teesha broke off, lips spreading in a secretive smile, as if she'd thought of a dirty joke she was almost too embarrassed to share. "Look, you know what? Forget the contract. What I'm saying, Jo, we're all in this together. If something's gone wrong for that sweet little boy, that's on *all* of us. You want me to go up there and talk to him, say the word—"

Jo shook her head. "Don't take this the wrong way. It's not me. It's Charlie. You really have to understand Charlie."

"Well, if you feel that way, you can't blame *us* for being concerned. He's been up there all morning, doing—but you know what he's doing. If we're going to get through this, as a family, then—"

"Oh, for fuck's sake." Sun Min clacked down her cappuccino. "Look. You may not care what the contract says. But I do. We paid enough to have them write the damn thing, exactly for situations like this. When it comes to the home environment, we *all* get a vote. And you've been outvoted, Jo. You just have to deal."

Teri and Aya both opened their mouths. Before anyone could talk, the lights winked off, the dynawindow blanked, and every counter, clock, display, and status window began to twinkle with inscrutable symbols. Chairs scraped, dishes clinked. The moms sucked hisses of surprise through their teeth as the bowels of the house thudded with mysterious operations. The ceiling projector whirred. Light returned, purplish, unsteady, as the walls filled with flickering images. Maybe mouths, maybe eyes, maybe fleeting faces, they reminded Jo, above all else, of rapidly gesturing hands.

*Not now*, she thought. *God, please, not now.*

"Sparklybits," Jo said out loud, "this really isn't the best time."

Strokes of brightness lit up the dynawindow, slashing upward, bending outward, like hands uplifted in a shrug.

Jo tried to remember the lingo. She thumbed on her sema and sketched a sign in the air, a series of slashes and chops that she hoped meant "cut it out." The light throbbed. The ovenex displayed a series of dots and carets, an emoticon row of blinking eyes.

"I mean it, Sparkly." Gesturing as she talked, Jo repeated the sign for *stop*. The dynawindow blinked. The walls displayed a now-familiar icon, an open circle with dashes on the sides, which Jo figured was supposed to resemble a bowed head.

"That's right, Sparkly. This is adult-people business. You go back up and wait with Charlie."

Fizzle. Wink. There were no words for the series of icons that glimmered and faded in the walls. When they were gone, the lights came back, the ovenex showed its default display, the dynawindow reverted to the rainbow tiles of the scheduler. There was no sound throughout the house except the swish of a bathroom-cleaning subroutine.

"Now that," said Sun Min, "is *exactly* what we're talking about."

With the ghost gone, the moms twitched into motion. Hands lifted to tuck hair behind ears, brush crumbs from slacks, adjust rings and necklaces.

"Sparklybits," Teri said.

"Really." Aya sighed. "You *named* it?"

"Well," Jo said, "Charlie did."

"And you let him?"

The tone of Aya's voice let Jo know this was about more, much more, than a simple name. It was about major failures of smart-home management, serious lapses of parental discipline, epic errors of motherly judgment. The frustration of a world-bestriding CEO at seeing a job badly done.

What could Jo say? What could anyone say? It was a pretty big deal, after all, letting the house get haunted.

Teesha gathered up Jo's plate and mug and brought them to the dish slot, ignoring the kitchen whiz as it scrambled at her heels. With the mess dispatched, she turned, drying her hands, and locked eyes with Jo, saying with a little nod, "I think it's time."

THE STAIRS TO CHARLIE'S ROOM WERE AT THE BACK OF THE HOUSE, BETWEEN THE self-care parlor and the door to the village commons. It was a screen-free zone: once you stepped into the hall, all devices automatically went on lockdown. Something they'd voted on ages ago. For purposes of stress management.

Yeah. Like that had worked.

The whole wing was tweaked-out for sensory modulation. Carpets, warm colors, floor-level lights, even a few pieces of hotelish wall art. The style reminded Jo of her mom's house, all the inert clutter and decor people had brought into their lives back then. As if they felt some anticipatory lack, a need to make up for the absence of technology.

The homeschooling room was at the end of the hall. Privacy checks clustered round the door: intercom, peephole camera, the pingpanel. All pretty silly, given that they'd equipped the place with round-the-clock monitoring. How many times had Jo logged in at the talkshow hour, zoomed in on nightvue to watch her son sleep? His face so placid as it dreamed, free of anxieties, until she could almost forget what went on during the day.

She slipped down the hall, aware of Teesha's heavy stride behind her. The other moms were still downstairs—scared, Jo supposed, of spooks and little boys. She lifted a finger.

Hesitated.

"It won't get any easier," Teesha said.

"I just . . ." Jo turned. Performance, she reminded herself. This was all a performance. "I feel like I already screwed this up."

Teesha's smile softened. "We all feel that way. Hell, we *did* screw it up. But we're fixing it now."

"Yeah, but I'm the live-in, you know? The one who's here. I don't want to say I feel closer to Charlie—"

The smile disappeared.

"—but I feel like I should have been on *top* of this," Jo hurried on. "As the person who actually, you know, stays in the house."

Bing: the smile came back. "We all had access to the logs." Teesha touched her hand. "We see the status reports. Any of us could've punched up the module, run a diagnostic. Hell, I *knew* the house had viruses. I just didn't know it was—I mean, I don't think any of us understood—"

"How attached Charlie was to it?"

"How bad things had gotten." Teesha's mouth pulled down in an expression Jo often saw on TV, an empathetic frown, acknowledging profound, shared wellsprings of emotion. "Know what I used to do? When he was little?"

"What's that?"

"I'd be in meetings, okay? Back when we were steering Senator Ramirez through his hot-mike hack. Sitdowns with the campaign manager, fundraisers, everyone completely losing their shit. So I'd have my specs on, y'know, scrolling feeds, saying I was keeping up with the reaction. Meanwhile, I'd have your updates on the periph. Diaper blowout. Major bed puke. Mystery crash at 3:00 a.m. Cake-face takes the trophy. All day."

"No!"

"A-yuh."

"Naughty mommy."

"It got me through. Don't think I coulda gotten through without it."

"Of course, most of those updates were the nanny."

"Sure. But you were *here*, Jo. That's what I'm saying. While I was taking care of grownup babies on the Mall, and Sun Min was cleaning up after authors in New York, and Aya was off selling her arm-spanx thingies, and Teri was doing... whatever Teri does..."

"Yeah, what exactly *does* Teri do?"

"God knows." They giggled together, and when the giggles subsided, Teesha touched her arm. "But you, Jo, you were right down the hall. Now, I know sometimes it might feel weird, being the one who—well, who can't quite pay in at the same rate. Which is fine. I don't think there's anything wrong with that. But when you have four professionals all butting heads—"

"I am a professional, though."

"Sure, you've got your nursing job. I think that's great."

"It's a licensed profession."

"It's fantastic that you do that. But you know what I'm saying. We're all equals here. Sure, Aya can be a big mamabear about nutrition. Teri's a hardass when it comes to finances. Sun Min's got a lock on the educational stuff. I'm sure I can be a little intense about all sorts of things. It probably feels like we're always on your case, like the homemaker always comes last."

"But I'm not—"

Teesha held up a hand. "You're valued, okay? No one thinks any less of your contribution. That's the point of a co-op, right? Everyone pays in whatever they can."

Sure, Jo was thinking, only some people pay in by hiring expensive online tutors, while other people pay in by screaming and bleeding in a hospital for ten hours. But she only smiled.

"We're power moms, right?" Teesha gave her a sock in the arm. "Wouldn't be doing this if we didn't like the challenge. That's what I tell my babies in D.C. Y'all think prepping for a primary is hard, try prepping a challenged kid for top-tier preschools."

Jo tried hard to keep her smile in place. "You ever have that feeling," she said, "like with every little thing we do, we're potentially fucking someone up for life?"

Teesha boomed a laugh, throwing back her head. Trying, Jo thought, just a little too hard. "Come on, now, girl. Into the dragon's den." She guided Jo's hand to the pingpanel. And pressed.

"Mom?" It always felt weird to hear Charlie on the intercom, how shrill his voice sounded, like something might be wrong. Jo leaned in.

"Hey, Charlie."

"Everything okay?"

Jo glanced at Teesha, got a nod of reassurance. "Sure, we just—can we come in? There's something we need to talk about."

A pause. Sometimes the no-screen thing drove her crazy.

"Thought you were doing your meeting out there?"

"This is part of the meeting. Something we need to discuss."

"Okay, but I'm doing my puzzles right now. I'll schedule you for . . . fifteen minutes."

Jo broke the ping. Teesha reached for the door, but Jo pointed at the status display. Purchased from a trendy tactile-play site, it was a custom-order birch flip-card machine, all-natural, nondigital, except for the gizmo that changed the letters. With a flutter and click, it switched to the phrase *Mom Time*.

"Cute," Teesha said, as Jo opened the door.

The homeschooling room continued the analog theme: sound-dampening rugs, a plain glass window. The only piece of modern tech was the learning station itself, a topline Sony schoolbox, with apps and doodads up the wazoo. Parentally controlled, of course, though not, apparently, controlled enough. This was where Charlie spent all his time, at least until the day when they'd finally cave in and get him a palmcom, at which time, Jo figured, he'd be lost to them forever.

He didn't like to use the headset, but plunked himself down at the screen and—zonk—went straight into Charliespace. For a kid like Charlie, Jo sensed, all of reality was basically virtual. When he heard Jo coming, he said without looking up, "We did the Parthenon and the Colosseum and Notre Dame and the statues. Except Sparklybits started painting the statues, but I told him that wasn't always historically at-at-attested," he finished with a dip of his head, forcing out the big word.

Jo started at a tap on her shoulder. "I'll get the others," Teesha hissed, vanishing on a whisper of carpet.

"Except we might do the color later," Charlie said. "But I told Sparklybits I have to check sources."

"Mmmm." Jo went to stand beside him. When she let her fingers trail in his hair, Charlie looked up in mild surprise. For Charlie, Jo had realized, the existence of other people was always mildly surprising.

"Then we're going to—" He broke off, moving his hands in jerky patterns, faster as frustration mounted, until Jo nodded and smiled, mimicking his signs. Charlie smiled, too, relieved at not having had to use words.

A babble from the hall announced the arrival of the moms. Teri descended first, nails flashing, keening with exaggerated joy, "Charleeeee!" Aya came next and tousled his hair. Sun Min nuzzled him. Teesha crushed him. Jo cringed, knowing how Charlie would feel about all this. But he handled it well, putting on his visitor face, as each mom claimed her obligatory hug.

"We're—" He pointed at the learning station, waving his hands in expressive swirls as his face bunched up. "We're—"

"Slow, baby," Teri said.

"Use your words."

"We're doing—"

"Take your time, Charlie." Teesha knelt on the carpet, nodding to coax him along.

"We're doing—"

"Puzzles," Jo translated, mirroring his gesture, and saw his face relax.

"Puzzles," Teri said through her TV smile. "That sounds so fun! Can I do a piece?" She scanned the floor, the shelves.

"No, they're—" Charlie made the sign for the learning station, then the sign for online, then a bunch of signs that meant something like "shapes of light." That was as much as Jo could follow. What he called "puzzles," she would have called "models": digital constructions, meant to be educational, that challenged kids to assemble famous buildings. Charlie had built the whole set so many times that Jo couldn't imagine he got anything out of it except the satisfaction of the process itself. Tick-tick-tick, piece by piece, a series of gestures he could have done by now in his sleep.

"Me and—Sparklybits—"

He was almost totally signing, now. Those wide, florid gestures that only Charlie fully understood. Charlie, that is, and one other entity.

At the name Sparklybits, the moms all turned to Jo.

"I see." Teri's smile was stapled on. "And how much time do you spend, Charlie, playing video games with Sparklybits?"

Charlie frowned. "They're not really—"

"They're like virtual building sets," Jo explained.

"Ah."

"Charlie. Sweetie." Aya knelt. Too close, Jo thought, but restrained herself. "Can I see how you play with Sparklybits?"

Charlie looked up for approval. Jo nodded. Sun Min and Teesha and Teri were already herding her out of the room, down the hall. Over Sun

Min's shoulder, Jo could see Charlie gesturing, flamboyant with frustration, as Aya cocked her head and tried to smile.

"This, *this* is what we're talking about," Sun Min hissed. "Right there."

"Does he always express himself to you with those… movements?" Teesha asked.

"Well," Jo sighed, "mostly with the ghost. But I've picked up some of it."

"This is why he's lagging." Sun Min jerked a hand at the bedroom, folding her arm like an Egyptian painting. The resemblance to Charlie's gesturing was uncanny, but Jo held the thought. "When did he start to miss milestones? I'll bet it was right when that thing showed up."

"It's more than a speech lag," Teesha said. "He's behind on every track."

"It's all connected. Speech, socialization. This is why his metrics have crashed. How can he succeed if he can't even talk?"

"He does okay on the homeschool stuff," Jo said.

Sun Min used the same face she probably pulled on authors who pitched digital-addiction memoirs. "You're going to fix an emotional lag with homeschooling? This is why we're pumping in 20 percent for Artemis Academy. Trust me, I've read the lit. You cannot, can *not* hit benchmarks across the social skillset without at least fifty per week of face-to-face time. Where are his public speaking skills? His prosocials? Empathy, engagement, emotional literacy? Do you know how badly he's lagging his cohort?"

Lagging. If there was one word Jo could've x-ed out of the discourse, it was that deeply loaded word, *lagging*. When she was a kid, people had said "catching up"—as in, "Jo Clark is still catching up in math." Before that, the favored term had been "behind." Go far enough back, people had used words like "retarded." The whole idea being that everyone was on the same road, all heading to the exact same place.

"How is he going to get into college with these benchmarks?" Sun Min threw up her hands. "Not just a good college. *Any* college."

"Maybe he won't want to go to college," Jo said, and knew instantly that she'd blown up the conversation, gone straight for the nuclear option. She might as well have hauled down her jeans and pissed on somebody's sandals.

"You're seriously planning to keep our son out of college?"

"I'm not trying to keep him out, Teri, I—"

"You're going to ruin his future, *our* future, because you don't have what it takes to run a household?"

"Teri." Teesha put a hand out.

"Because you let this piece of rogue code get into his brain and—"

"Teri, Teri, Teri." They were all pressing round, trying to calm her. Something had slipped in Teri, the newscaster composure an all-or-nothing proposition, now firmly jammed in the off position.

"Thing is, Jo, *you're* not paying for the private schooling. *You're* not paying for the prep, therapy, emotional tutoring, nutritional advising. Twenty percent of your salary? I'm sorry, that's a fucking rounding error. The rest of us? Okay?" Teri's fingernail scrawled circles overhead. "We're the ones working seventy-hour weeks, traveling the world, just to pay for this goddamn regimen. Why? Because we want the best for Charlie. Because motherhood, call me crazy, I happen to think that's kind of an important job. But you—"

"All right, Teri." Teesha held her arm, but Teri threw her off with a clash of bracelets.

"What am I going to tell Frank?" She thudded partway down the stairs, looking up at them, tapping at the tears on her cheeks. "He already thinks this is some vanity trip for me, like adopting a fucking gorilla at the zoo. Now I'm supposed to ask him to be the dad of a budding high school dropout? Jesus, I *should* have adopted the gorilla; at least they don't grow up to join Machinima porn fandoms. 'Cause I'll tell you, that's where this kid is headed. Every other mom in the office is beating the benchmarks. Every one. There's a single father in technical support who's got a son in the ninety-eighth bracket. And we've got a lagger. God."

In the silence that followed, a funny sound came from Charlie's room. Jo took it at first for a technical failure—audio feedback, a broken speaker—until she pegged it as Aya's cry of surprise.

"It's not Charlie's fault." Sun Min looking at Teri with what might have been sympathy or distaste. "It's—"

But now Aya was hurrying to join them, bustling up with her brisk executive stride, planting hands on her waist to announce, "It's happening."

"You get through?" Teesha asked.

"To our kid? No. But there's this." Aya semaphored at the wall, remembered the no-screen thing, yanked out her phone and tapped into the house system, swinging it to show everyone the feed.

"Shit," Jo said.

•••

They were in the kitchen, gathered around the dynawindow, watching the feed from the village gate.

"How long has he been waiting out there?" Jo asked.

"I texted him to stand by." Aya clicked in for a close-up. "Yeesh," she winced, "just look at him."

Jo had to admit that the man at the gate wasn't especially prepossessing. Particularly not on the zoomed-in security feed, with its anti-blur motion compensation and refractory enhancement and high-def-whatever and all the other brilliant tweaks the geniuses of home security had put into the software. There were slews of apps to make people look good on video; this particular program revealed them at their worst.

Not that the man at the gate would ever have looked especially good. When Jo met him, he'd had a kind of greaser-trying-to-clean-up-his-act vibe, hair slicked back, a flush in his cheeks, like a slacker who'd just stepped out of the shower. On-screen, now, he looked like a hair malfunction at the hippie factory. Like someone had swept up Chewbacca's haircut clippings and glued them to a giant peeled potato. There was no good stage of life at which to have that look, Jo thought, but thirty-six—which is what she guessed the exterminator was—was too old for any conceivable excuse.

"You checked his background? Consumer reviews? Credentials?"

"I met with him. He showed me his shop."

The man began to pinch his nose, pulling hard to squeeze the boogers out. The kind of semi-discreet nosepickery you might get away with on a busy train, but definitely not on close-up video.

"His shop, huh? He have any dead bodies there?" Aya sighed. "Well, let's get this over with." She punched the code for the gate and the guy slouched in, schlumping through the New Urban street plan with his truck moseying along behind him.

As he came around the corner to their street, the moms all trooped outside. "Hello ladies!" the exterminator yodeled up. The truck, still creeping at his heels, gave a beep. "Park!" he commanded it, pointing to the curb, and jogged up the steps to loom over them in all his ungroomed glory. "So." His teeth peeked through a wickerwork of hairs. "This is the coven, huh? I'm Evan."

"Let's go over the situation." Aya spun on a scraping heel. "Then we'll tell you how we want to proceed."

Evan bumped his head on the doorframe coming in. He seemed not to notice. He bashed his shoulder on the turn into the hall, seeming not to notice that, either. He was looking at the ceiling.

"Yeah, you see a lot of hauntings in these older units. Legacy wiring. Puts a limit on your hardware. So people don't push the updates, a backdoor opens, and 'fore you know it, boom. Spooktown. What's the interface here? Still got the old semaphore hookup?" He stuck a sema on his thumbnail and signed the lights-out sign, snapping the hall into darkness, occasioning several bumps and curses. "Nice."

"Jo tells us you're quite the expert in these matters," Aya said, leading the group into the kitchen. "Credentialed," she added, with a hitch of her eyebrows.

Evan shrugged. "Expert, ah, that's not really the word I'd use. Freakishly obsessed is more like it." He went to the ovenex and started poking buttons. "Honestly, this stuff is like the only thing I ever think about."

"N-o-o-o-o." Aya looked him over. "Can it be true?"

"My pops, he was *way* into rogue AI. Used to hunt 'em, all through the hotlands. That was my first childhood memory. Rolling with a pack of ghostchasers in Lou'siana. Course, these days it's a lot easier. There's people who'll just sit around and wait, rig up some bait, and hope the things'll show."

"And what works as bait for a ghost?" Aya asked.

"Well, it sounds awful, but, the truth is: kids." Evan pulled out a crumpled pack of gummies and popped some in his mouth, chewing with much bristling of hair and smacking of lips. He opened the trash panel, peeked inside, went to the drone dock, and flipped a switch. "Not literal kids. But the stuff a kid'll do. Poking around. Punching buttons. Messing with stuff. Come on, Sparkybits, where are you buddy?"

"Sparklybits," Sun Min corrected, and winced at her own complicity.

"Yeah, 'cause y'know, they're basically kids themselves." Evan noticed the way everyone was looking at him. "I mean, not really," he clarified. "They're just software. But *as* software, they're still learning the ropes." He went to the dynawindow and waved his arms. "Got a shy one here, huh?"

"It mostly comes out for Charlie," Jo said.

"Gotcha."

"So this is common?" Teri's voice still hadn't quite climbed down from the pitch it had reached earlier. "A kid, connecting with one of these things? That's normal?"

"*Normal* is not a word I really like to use." Evan brought up the security panel, tapping monitors until he got Charlie's room. They looked at the boy's bent back and head, hunched in front of the learning station. Evan's fingers vanished into his beard. "They were built to be learners. You know? Pattern matchers. See, we think of learning like a thing that happens when you're taught, right? But it's more like a thing that just plain happens. And learning with someone else, I guess that's easier than learning alone."

The puzzles were back on Charlie's screen—even on the feed, Jo could make out the shapes—and the boy had begun his eloquent gesturing, tracing loops and swirls in the air. Evan seemed to be unconsciously mimicking him, swinging hands and fingers, until he abruptly stopped and scratched his chin.

"Guess we better go up there," he said.

UNDER NORMAL CIRCUMSTANCES, THE DOOR TO CHARLIE'S ROOM WAS KEPT CLOSED, but the visit from the moms had thrown off the routine. As they went upstairs, they had a clear view of his hands, waving in front of the learning station screen. They could see exactly what those hands were doing.

"Whoa." Evan pulled the kind of face that emojis could never capture: mouth screwed up, eyes slightly out of focus, eyebrows riding high above a grimace that seemed to say, *I don't know what you make of this all, but damn, what a show!*

In his pod-chair, Charlie was slumped in classic kid-posture—a sprawl of boneless lethargy, except for his arms. These were animate as tentacles, weaving, swishing, fingers wriggling like strange sea creatures, plucking invisible meanings from the air. Here was the performance to which his earlier fumblings had been a kind of rude prelude. Anyone could tell the gestures constituted a language—just not a human one.

The patterns in the screen were eerily similar: curls, twists, and ribbonings of color, animate icons of light.

"Good God," Aya breathed.

"That's no home semaphore," said Evan. "No, sir."

It wasn't ASL either, Jo knew, or any other human sign system. She'd checked. It was a language, she suspected, that had never been used anywhere outside this house.

She became aware of a stumbling pressure, a bumping hip, a foot on her toe. Teesha was railroading the whole group down the hall. In the bedroom, Teesha eased the doorway partway shut, peeking out into the hall. "Can Charlie hear us in here?"

Jo sat on the bed. "He wouldn't notice if he did."

"Now that," Evan pointed down the hall, "is quality specter-speak."

"But is it—" Sun Min restarted her question. "*What* is it?"

"It's how they talk. Mostly. Though not usually at that level. By which I mean, well—" Evan socked his tongue into his cheek. "Is Charlie, uh, special?"

"In what way?"

"You know, gifted?"

"He used to be," Sun Min said.

"A smart kid?"

"Charlie is . . . mathy."

"Focused."

"On the spectrum."

They all had their own terms for it, picked up in parent gossip, office chatter, the world of online mom-chats.

"Charlie," Jo said, "he latches onto things."

"I mean, the kid's a nerd. Right?" Teri shrugged. "We wanted a nerd. It's what we paid for. A boy who'd ace the tests."

"And not a girl?" Evan's question evoked a chorus of half-hearted mumbles.

"It's a bump," Jo explained. "Having a boy. The top-ranked colleges are 65 percent women. The top-paid professions are 65 percent men. Think about it."

"Okay," Evan said noncommittally, puffing his cheeks.

"What did you mean, it's how they talk?" Sun Min narrowed her eyes. For the past few minutes she'd been compulsively clicking the clasp of her handbag. "It's like a code?"

"Well, technically it's all code." Evan grinned as they groaned. "I mean, the ghosts, they came out of the omnicom craze, right? So, like, everything has a mind, okay? Your shirt, your coffeemaker, your car. Well, what kind of stuff is a shirt gonna talk about? If it talks to a coffeemaker, what're *they* gonna talk about? If all those things are talking to each other—"

"But they still have to talk to people."

"Sure. But the thing about smart devices, they're mostly talking to other devices, *about* people. The whole point of the internet of things is it's networked. So if you're a free-floatin' free-lovin' higher-level entity that grew out of all those little programs—well, how the world looks to you, it's probably mostly not about human language. Like, User72 is heading southwest at sixty miles per hour on Highway 92, elevated blood pressure, restless, making hungry faces, scanning the map.... You put that all on a screen, what do you get? They can use some English, sure. But they don't *think* like we do. So they don't *talk* like we do, either."

"But they don't think at all." Aya's voice had a calculated coolness, the tone she probably used to close agenda items in meetings. "So what makes anyone think they're talking at all?"

Evan looked at his toes. To Jo, at that moment, he looked just like Charlie, getting grilled on his social skills in some therapist's office. He brought up his head with sigh. "Look, you've got a couple of options here—"

"We want to get rid of it." Aya cut him off.

"Okay, but listen—"

"No. We want to get rid of it."

"Can't we just chase it off?" Sun Min was still fussing with her handbag. "Get it out of the house, but not, you know, kill it?"

"Well, you can do that. But they usually come back. Creatures of routine, right? Once they bond with a child—I mean, once they've linked to a user—"

"I don't understand." Teesha swung an arm to break into the conversation. "It's software. Ones and ohs. If we try to delete it, can't it copy itself?"

"Sure. They don't really like to do that, though."

"Huh?"

"Yeah." Evan's inner geek broke through in a goofy smile. "Well, it's a funny thing. The thinking is, if you're nothing but a trail of bits, copying is like a form of movement, right? Imagine you left a clone of yourself every time you took a step. Well, to a program, what we think of as movement is basically a special case of copying, except you delete the original version. That's part of what makes these things so special. A virus copies. A ghost *moves*."

"So if we delete it, you're saying, it's gone."

"If we can manage to delete it, yeah."

The words had a solemn power—the power, Jo suspected, of any statement of finality. It took her a while to realize everyone was looking at her. Jo made herself focus on Evan. She had a vision of Charlie all grown up, stranded in some weird niche job, letting his hair run riot, dead-ending his life. Becoming this man.

"It's like we discussed, Jo," Evan said.

"Right," Jo sighed. "Like we discussed." She got up and said, "Let's do it, then." And headed down the hall to Charlie's room.

HE WAS AT THE LEARNING STATION, STILL SIGNING WITH SPARKLYBITS, PERFORMING those strange, lavish gestures. Jo thought of Evan's explanation, how a coffeemaker might talk to a car, a car to a faucet, a faucet to a chair, but none of that struck her as having much to do with language. Language was about empathy, expression. Sharing something other than information.

"Charlie." She slid into his line of sight. He registered her presence, blinked, broke from the screen, and gave a mind-clearing shake of his head. The human child slowly came back into his face.

"We're doing a new puzzle," he said.

"That's great. But Charlie, listen, I need you to put away the puzzles for now. There's something important we have to talk about."

His eyes became deerlike, anticipating a shock. Did he know what was coming? Not a chance, Jo thought. There would have been tears, shouts, a crisis.

"This man," she said, "needs to talk to Sparklybits."

"No," Charlie whispered, so faint Jo was sure no one else had heard. She tried to ignore the lance of ice in her heart.

"Hey, there, Charlie!" Evan used that awful adult-talking-to-a-kid voice, lowering his bulk to the floor with a grunt. Charlie glanced over, noted his existence, and turned back to Jo, saying only to her, "Mom, please."

"Sparklybits, Charlie, he has to be . . ." Jo couldn't finish.

"It's okay, man." Evan rocked backward, wriggling a hand into his shorts pocket. "We just need to keep him from running around loose. Catch him and put him in a safe place, you know?"

Jo winced. Did they really want to tell him that? Before she could send a signal to Evan, though, Teri picked up the theme. "That's right, Charlie. We need to make sure he's safe."

"It's for the best," Aya said.

"For your health," said Teri.

"There are other ghosts," Sun Min added. "All different kinds. Different types of AIs. Right?" To Evan.

"Oh, sure," Evan soothed, "all kinds. I mean, I even have a bunch. At home."

Again, Jo made eyes at him, but Teesha was talking.

"And all kinds of other friends. Real friends. Wouldn't it be nice, Charlie, to have some human friends?"

"We'll just snatch him right up and make a nice home for him." Evan pulled out a gizmo and plugged it into the learning station, making various IT-dude adjustments. "To do that, though, Charlie, we need to know where he is. And to do that, we're going to need your help." Looking up from his gear, he mouthed over Charlie's head, *Ready*.

Charlie was still staring at Jo. She knew what he wanted from her. Not reassurance, not explanations, but someone to cut through all the placating bullshit, tell him how things really were. Curling her fingers around his hand, pushing down her emotions to make room for his, Jo said, "We have to do this, Puppa. I should have told you sooner. I'm sorry."

"No." He mouthed the word, then howled it. "N-o-o-o-o!" Before Jo could react, they were in full meltdown mode. Charlie grabbed her arm as if to claw his way up it, into her head where he could change her mind. "No, no, no!" Jo avoided looking at the others. Charlie really did seem, right now, like a much younger child, emotionally stunted, behind the curve. Lagging, undeniably lagging, as he vented his grief in a series of screams.

"I'm sorry," Jo said, struggling to hold him. "Puppa, I'm sorry." But every apology, she knew, was a sentence handed down, a judgment, a verdict, a punishment. As if his sorrow had exceeded the reach of human speech, Charlie jerked away and made a sweeping gesture, lifting and thrusting out his fists.

"Charlie!" Aya gasped, mistaking it for an act of aggression. But the only violence here was the violence of passion. Charlie's fists opened into a gesture of loss, gathering in toward his chest and flinging outward, as if hurling clusters of invisible blossoms, expressing a sentiment for which English had no words. It struck Jo as curiously archaic, elemental, like something from an opera or pagan ritual, a display of mourning the modern world had lost. She wondered how the ghost would express its distinctive digital experience, looking back through the networks of the world and seeing a million lost copies of itself.

As Charlie continued his ballet of supplication, the lights fluttered, the walls groaned, the screen of the learning station began to swirl. Streaks of light curved down and inward, forming gentle cupping lines. Sparklybits had come to see what was the matter, concerned for its suffering human companion. Poking its cyber-nose, like an animal, right into their trap.

Without a sound, without any obvious signal, the room subtly changed, becoming stiller, steadier, as the screen of the learning station blanked, leaving only a few blocky pieces of the puzzle that Charlie had been building with his friend.

"Got him," Evan said.

Dropping his arms, Charlie sank to the carpet, wrapping his arms around his head. Teri stroked his hair. Sun Min murmured explanations. Teesha bustled up with grandmotherly authority. But Charlie dragged his pod-chair to the corner and sat staring at the empty wall. And Jo couldn't help feeling, even though everyone gathered around to apologize, that this last gesture of rejection had been meant entirely for her.

NO ONE SPOKE AS THEY WENT OUTSIDE AND STOOD SHUFFLING THEIR FEET ON THE concrete steps, all somehow avoiding, by one shared instinct, the temptation to glance back into the house. Charlie followed them, but there was no forgiveness in this act; Jo knew it was only a concession to routine. As they squinted into the reddening sun, Evan jogged down the steps, moving in the springy sideways trot that Jo associated with more athletic men. On the path he looked up, he shaded his eyes.

"Well, if it ever happens again . . . you know who to call." Evan hesitated, seeming to feel something more was needed. Then he cocked a finger at the house, squinted one eye, and said, "Zap."

Don't overdo it, Jo thought, glaring down. Evan swung his arm in what was probably supposed to be a bow. "Luh-*ay*-dies." A moment later they were watching his truck putter away.

"Well, Charlie." Teri squatted, shining her camera-ready smile into his face. "I'm so sorry we didn't get to catch up more. I love love love to see you, sweetie."

"Yes, so much," murmured the other moms.

"And you know, Charlie," Sun Min hesitated, maybe second-guessing what she'd been about to say, but plowing ahead anyway, "it really was the right decision."

"Yes." Aya nodded. "For your future."

"For the family," Teesha said.

"Kiss goodbye?" Teri squealed, flinging out her arms. The question usually won from Charlie a grudging hug. Today it received only agonizing silence. Hanging their heads like scolded children, the moms shuffled away down the path, holding key fobs aloft to let out a froglike chorus of peeps, summoning their rental cars from the village lot. They would already be rehearsing, Jo thought, the things they'd say to other parents at the office, to colleagues and boyfriends, to their own mothers at home—in Tokyo, in London, in airplanes, clubs, bars: "Mothering is hard." "It can break your heart." "Just be glad you don't have kids." Jo herself was wondering what she'd say at work, in response to the inevitable Monday morning questions.

The Zephyr had come out to wait in the drive. Charlie jerked open the rear door, flumped in, and slammed it. Jo got into the driver's seat, tapping a route into the console. They pooted out, following the golden trails of cyberspace, turned at the corner, and rumbled through the gate. On the highway, they took the second exit, bumping down to a strange part of town.

"Mom?" Charlie broke his vow of silence, leaning forward to put his face between the headrests. "Where we going?"

"No Doctor Brezler, today," Jo said over her shoulder. "We have, uh, another appointment." Without looking, she put a hand behind her ear, grazing his cheek with her knuckles. "Change of plans, Puppa."

If he guessed what she had in mind, he didn't let on. The car zagged through a series of turns, paused in front of a pizza parlor, recalibrating, then set off into a section of town that seemed to have been rezoned for random uses. People were squatting in public garages, selling scrap out of gutted franchises; an old YMCA had been refitted to house a group of refugees. Folks in the street sold vegetables, flags, homemade liquor. The car wriggled through a cluster of tents.

Only when they were a block away did Jo begin to recognize the area. The building itself was unremarkable, a strip mall in which most of the units had been converted to shabby apartments. The last shop, a former game parlor, still had a tangle of fluorescent tubes in the window, tracing the outlines of crossed pool cues. A hand-painted sign read "Ghostblasters!"

"Mom?"

Jo clucked for silence. If she'd learned anything from this ordeal, it was not to say too much, too soon.

The door jingled a welcome. No one stood at the dusty counter where a register had once been perched, no one guarded the fire-retardant curtains that blocked off most of the main floor. Jo pushed them apart. The place was smaller than she remembered, but jam-packed with the kind of interesting clutter that can make a room feel paradoxically large. Appliances, drives, peripherals, gadgets, all sprawled across the old, battered pool tables, linked by kelpy mats of wire. Looking them over, Jo was mostly conscious of a festive abundance of lights. Like candles, she thought. Like a birthday surprise.

"Boo." The sound actually made Jo jump. Evan popped from behind the nearest table, brandishing the palm-sized gizmo he'd brought to the house. He presented it with a flourish. "Madame? Your ghost."

Jo turned to Charlie, expecting—but she wasn't sure what she was expecting. He seemed not to have heard what Evan was saying. He was staring goggle-eyed at the wilderness of wires, this doll-size metropolis of tiny night fires. His hands lifted, clutching. Jo didn't need Sparklybits to tell her what the gesture meant.

"Like it?" Evan said.

"They're so—what are they?"

"Ghosts." Evan reached out, letting his hand fall on a gadget at random, a toaster, walking backward to get a closer look. "This one, let's see, this is old Elmo. Ancient feller, small memory, doesn't need a lotta space. I keep him here and let him ring the bell. Every once in a while I hook him up for some TV time. He likes that."

Charlie ambled along the aisles, lifting his feet as if by practice over the rubber strips laid over the roots of bundled cable, his eyes locking on to one gadget after another. Evan shambled behind him.

"Over here we have Skittles. One of our big vocal communicators. Talks in a tone-scale kind of like a whalesong. Probably appeared in a house that was blind-adapted, sound-heavy interfaces, that kinda thing. Had a real tight bond with this girl up in the estates. This here, this is Wanda. Kind of a retiring type, but she just *loves* chasing fingers on a touchscreen. The simularium, here? That's our condo. Whole ghost family packed inside."

"This . . . this is so *harsh*."

"I'm not up on the lingo, man, but I'll take that as a compliment. Have a look around. Tap the screens, touch buttons, whatever. Maybe break out some semaphores. They love the attention."

As Charlie worked his way through the gizmos—hesitant, at first, then with growing enthusiasm, and finally with invincible levels of absorption—Evan sidled up to Jo, whispering, "So. We're good?"

She lifted a shoulder. "Seems that way."

"You were right, then, huh? Day-yum. Whole thing went off like you said."

Jo nodded, not wanting to tell him how wrong he was, how far the day's events had diverged from her expectations. Evan wagged the device, its corkscrewed tail of cable flopping: the new home of one Sparklybits.

"How's that whole deal work, anyway? Like, those other ladies, they just chip in some money? Rent your kid for a weekend or something? I never understood the whole co-op family thing."

Jo kept silent. Evan must have seen in her face that he'd brought up a not-OK subject. "Well, anyway," he shrugged, "you were right about how they'd take it. I wonder what they'll think, if they find out you—"

But here came Charlie, rushing through the aisles, brimming over with syllables of delight, grabbing Jo's hand and dragging her away to share the discoveries of the last five minutes—as if, with the fluid enthusiasms of childhood, he'd already forgotten his earlier grudge. She had to tell him three times before he noticed the gadget Evan was holding. Then it took three more tries to explain the thing's significance. Even after he understood what it meant, Charlie's reaction wasn't quite what Jo had expected. Almost with reluctance, he let Evan place the drive in his outstretched hands, solemnly closing his fingers around the plastic, stretching out a finger to stroke the screen. A streak of light appeared and faded: a glimmer of *Hello.*

Jo didn't have to remind him to say thank you.

"HE'S GOING TO HAVE TO STAY IN THERE. AND THERE ARE GOING BE SOME RULES. NO use in the house, for one. Or during school. This'll be a special occasion kind of thing, not an all-the-time thing. Got it?"

Charlie squinted across the seat, then back at the block of plastic in his hand. When Jo nudged him, he looked up, blinking.

"Puppa, this is important. I know how it sounds. But we can't let the other moms know, okay? Not for now. At some point, maybe, when things have changed..." Jo decided this particular conversation could wait for another day. "The important thing is—"

"Mom?" He was looking at her with a face he wore often these days, an expression that scared Jo as much as it delighted her. It reminded her of the father she'd lost, of the husband she'd once imagined she'd have—of the man her son was slowly becoming. A smile that would almost have been cruel, if Charlie had been aware of what it did to her. "Thanks."

Jo waited until she could trust herself to speak. "I should have told you," she said. "About what we were planning. But... I wasn't sure you'd understand. Or that we'd be able to pull it off. Or I thought things would get messed up somehow, or that we—oh, I don't know, I just should have—"

They bumped into the drive, the Zephyr purring, waiting for them to hurry up and leave so it could enter the garage and do its nightly diagnostics. Charlie was fiddling with the gizmo in his lap, swiping symbols into the screen, changing settings, as he pulled a roll screen from the glovebox and pried back the rubber socket protector.

"I never got to show you. What we were building."

"I—" Jo took a second to recalibrate. "You mean your new model? That thing's not supposed to have any Wi-Fi—"

"No, no, it's okay. Sparklybits'll remember." Charlie unrolled the screen across his lap. "You really want to get the full effect."

The images were forming already, swoops of color, curving lines, sketchy shapes that gathered slowly, clicking together to form a blocky frame.

"Interesting," Jo said. "Is it a castle?"

"Kind of." Charlie gave a little smile.

"A palace? A fortress?" Jo angled her head as the pieces accumulated. "Is it a church?"

Charlie didn't answer. He'd begun to stroke the screen along with the ghost, adjusting, guiding, adding and deleting, making subtle edits to the spectral assemblage, contributing to the dance of shapes.

"A school?" Jo said. "A hospital?" Surprising herself, she made her own contribution, reaching down to trace the ghostly movements, letting out a laugh of surprise as the hovering blocks ticked into position. Charlie laughed, too, moving his hands more quickly, now—in loops, in jabs, in pirouettes of dexterous motion—and Jo sat back to admire his fluency, his eloquence, in this language with only two speakers, this culture of two souls.

Her eye drifted to the windows, the gold and violet shapes of dusk, and in a blink she had it.

"It's our house. Right? That's what it is. You're building our house!"

Charlie was silent, absorbed in his craft. Only when the work was almost finished did he look up, conspiratorial, grin slowly widening, as the details continued to accrete beneath this hands—the bricks, the fixtures, the dollhouse doors and windows—and a plush sweep of lawn where two tiny figures stood, joined at the hands, like ornaments on the phantasmal grass.

"Just wait," her son told her. "Just wait and see."

**Mary Robinette Kowal**

STEPPING OUT OF THE LIGHT RAIL ON WEST END AVENUE, GAIL FELT NASHVILLE'S humidity smack her in the face. It was like stepping into a sauna and being slapped with a hot, wet towel. Sweat slicked her skin. The space between her palm and her service eDawg's support handle became its own fetid swamp.

Wilbur mistook her hesitation for a Parkinsonian freeze and paused next to her.

"Wilbur, heel." Gail braced herself for the walk across Millennium Park to the Parthenon, the world's only full-scale replica of the ancient monument. And today, the site of yet another battle with the park administrator. Barnum's email still filled her veins with a fury to match the day's heat. *I stopped in to see how it was going. These are smaller than expected. We need to make plans to compensate.*

Compensate. It was an exhibition of Persian miniatures. The paintings were *supposed* to be small. When she got there, she would explain, again, that—

Her feet stuck.

Immediately, her faithful eDawg compensated, balancing her suddenly off-balanced body as Parkinson's nailed her feet to the pavement. Gail tried to step forward, but her feet wouldn't lift from the ground. All she got for her efforts were some half-hearted knee movements that didn't even get her heels off the ground.

Gail stopped trying. Agitation and multitasking made her symptoms worse. She took a slow breath, looking at the Parthenon sitting on its low, green hill with a wall of storm clouds piled behind it.

She wouldn't have the fight with Barnum until she saw him.

A message from the medication pump in her side appeared on her heads-up display. "Rescue dose?"

"No." It would start a bout of dyskinesia, and she needed steady hands to do the matting. Her Deep Brain Stimulator kept the Parkinson's tremors under control, but the random dyskinetic motions were pure medication side effect.

eDawg understood. The purple artificial dog took one step in front of her and flashed a laser line in the grass as a visual cue. Her internal metronome might be stuck, but she could still step over or on things. eDawg's bright red laser gave her a visual cue.

Gail stepped over the line. "Right foot."

Did she feel like an idiot talking to herself? Not anymore. "Left foot."

Two big steps later, she unstuck and walked the rest of the way to the Parthenon as if she were any other old lady out walking her robot dog. And she was not alone. Mixed among the teens on their hover scooters and dads with their strollers were other elderly folks out for walks with their robot companion animals. Most of them were themed as dogs, but she also saw a robot pony and a robot emu. Sure, any of them could have been an automated walker, but those didn't look at her with sad eyes when she skipped an exercise session or wag a stumpy tail with delight when they went out.

Programming, yes. But it worked.

Across the lawn, she saw one of the women from poker night and raised her free hand to wave.

"Watch out!" Someone slammed into her right side. Both of them tumbled over eDawg, who tried to compensate but staggered sideways. They all went down in a flail of limbs and robot dog and hover scooter.

The teen scrambled to their knees first and swept a riot of violet locs back from their face. "Oh no! Oh—I'm ever so sorry." Their British accent marked them as a tourist. "Are you okay? I'm so sorry—I just. I'm so sorry. Are you okay?"

"I'm fine." Aside from an elbow that was screaming and what was sure to be a massive bruise on her left hip. Her service dog extricated himself from them and moved to Steady position in front of her. Gail put her hands on Wilbur's warm, smooth back like a bench. "I've had worse falls all on my own."

"I'm dreadfully sorry. I lost control of the scooter." The teen offered their hand to help her up the rest of the way. "I'm Keisha Brown. He/him."

Gail had noticed the loss of control, but gave the boy the kindness of not being sarcastic out loud. She took his hand "I'm Gail Krishnasami.

She/her." Between eDawg and the teenager, Gail got to her feet with grati-fying ease. "New to Nashville?"

He nodded, bending down to right the scooter. "I always see them on telly and wanted to try one. I'm so sorry. I was going too fast and..."

"It's all right." She gestured across the great lawn to the Parthenon on its low hill. "Have you been to the..."

"My mums are in there now with my kid sister." He shrugged, studying a scuff on the rental scooter. "I've seen the real one. This is just... concrete."

"But this is the only full-scale replica of how it used to—" Her phone rang in her ear, with her wife's alert signal. Of course. She would have gotten notification about the fall and always checked in right away. If Gail ignored the call, Bobbi would assume the worst. She grimaced and held up her hand. "Sorry—incoming."

Keisha nodded and wheeled the scooter around, walking it away. "Take care."

Sighing, Gail activated the connection. "Hi, Bobbi. I'm fine."

"You had another fall." Her wife's voice was tight with concern. "Are you all right?"

"I just told you I was fine." She slapped her left thigh, and Wilber moved into heel position. "A teenager just lost control of a hover scooter."

"I can see the GPS. How many times have you fallen on the way to work?"

"Someone running into me is not related to Parkinson's." She started walking, because otherwise, Bobbi would start asking if she were frozen. There were times when, as much as she loved her wife, Gail deeply regret-ted that she'd agreed to let Bobbi get fall alerts. Gail needed to derail this conversation before her wife really got going. "What are your plans for the day?"

"It's still a fall. The eDawg is supposed to prevent those and it isn't."

"I would have fallen a lot more often if not for Wilbur." She kept her gaze fixed on the Parthenon and activated Wilbur's laser prompt. If Bobbi kept on like this, Gail would absolutely wind up freezing out of sheer agitation.

"My point exactly. Honey, they're getting more frequent." In the pause, Gail could picture Bobbi rolling the end of her silver braid between her fingers. "I think we need to revisit the retirement conversation."

"We'll talk about this another time." If not for the GPS, Gail would lie and say that she was walking into the building now. When she'd started there fifteen years ago, the massive concrete walls would have cut off the call for her. Repeaters meant that it would continue without a bleep, and that did not seem like a positive in the moment. "I need to prepare for a meeting with my boss about the new exhibit."

"Gail..."

"Bobbi Ruth Varnell, so help me if you keep talking about this I will tie a skunk to your belt and wallop you from here to Wednesday." Gail blew breath out her nose like a racehorse. The Parthenon loomed over her. Beyond it, the sky had the green tinge of an impending thunderstorm. Maybe she'd get lucky and they'd lose power. "I have to go."

"Sweetheart—"

"I have to go. Walking into the building now. Bye!" She'd said goodbye, so it didn't really count as hanging up on her wife, but just by a thin split hair. Gail set her phone to Do Not Disturb and very deliberately turned off the GPS and fall alert. It would send an update to Bobbi that she'd deactivated it, and that would just have to be part of the larger conversation. At the moment, she had to focus on talking to her boss, and she needed to be reasonably calm for that.

Gail dodged tourists as she walked down the short tunnel leading under and into the Parthenon. When she pulled the door open, the air conditioning rolled out to grip her in a dry Arctic blast.

She sighed with relief as the sweat on her skin suddenly did its job and cooled her, instead of sitting there as another layer of misery.

From behind the welcome desk, Slim Jenkins lifted their exquisitely manicured hand and gave a gold-embellished wave. "Morning, Mx Krishnasami."

"Morning, Slim." She nodded toward the small gallery that filled the back half of the basement of the Parthenon. Brown paper covered the glass partition as they prepared for the new exhibit. "He in?"

The young person fiddled with their bolo tie. "He's... in a mood."

Barnum Smith was the youngest park director in the history of Nashville. He was good at his job, but occasionally had a case of Imposter Syndrome that he overcompensated for by micromanaging his staff.

"Bless his heart, so am I." She headed for the stairs, guiding eDawg around a gaggle of teenagers on a fieldtrip. They were paused at the cases of artifacts, tapping the air at the interactive digital presentations that had popped in their heads-up displays. That or playing Mindfox or GalaxyKitten or whatever the kids these days played on their HUDs.

The door unlocked as it recognized her and hissed open. Inside, Barnum was leaning over the table Gail had been working on. His spiky blue hair caught the light like the plumes on Pallas Athena's helmet. If only he had as much wisdom as enthusiastic vision.

"Gail. I want to talk to you about—"

"I got your email." She let go of Wilbur and the eDawg's handle retracted into his back. "And I'll gently remind you that I sent you the dimensions of the miniatures with my original proposal."

"Sure. But in context, they're too small. No one will be able to see them." He turned and looked at the walls. "We can do a projection on the end walls."

"They are designed to be small and to be viewed up close." She walked to the table, with her eDawg following as if he were glued to her. "Part of what makes these so amazing is that they were painted by unaided human hands at this scale. Enlarging them removes that wonder."

"It's an accessibility concern."

"We've got plans for people with vision challenges. The virtual display will open on their HUDs and for those who prefer analog, we have magnifying glasses." She leaned over the table. The top painting had a garden filled with intricate tiny flowers and dotted with gilt. In the top center, Anahita, the Persian incarnation of Wisdom, rode a chariot drawn by four horses.

She had built the entire exhibit around this one, rare depiction of Anahita in the miniature school from the classic period. The other miniatures showed scholars or women reading, in ways that related to wisdom, but were not an overt depiction of the goddess.

"I just think that it's going to show poorly against..." He waved his hand toward the ceiling. "That."

"That." Above them, in the Naos, a reproduction of Athena Parthenos stood at an impressive 12.75 meters. Gail straightened the paintings in their protective sleeves. "The contrast is the entire point. The exhibit is about the reverence of wisdom and that scale is immaterial. You liked this, specifically because the emphasis on miniatures would appeal to school children who always think they need to be big."

"Yes, well that was before I knew that they would be so small." Barnum grimaced and pointed to a painting of women listening to music near him. "I'm not denying that they are beautiful. I'm just saying that enlarging them for ease in viewing is a prudent choice."

"So you want the exhibit to make a visually semantic argument that something has to be large to be appreciated."

He rolled his eyes. "No. That's not what I'm saying. I'm saying that—"

Tornado sirens cut through the conversation. On Gail's HUD, a weather warning popped up and her AI's voice came out of Do Not Disturb. "Alert, alert, alert. Tornado warning. Seek shelter."

The HUD showed the tornado overlaid on a map of Nashville. It was heading straight toward the Parthenon.

Across from her, Barnum said, "Oh, shit."

How many people had she seen outside the Parthenon? All of them would have gotten the same alert with directions to the closest shelter. "People are going to be sprinting for us."

"Safest place." Barnum headed for the doors of the gallery. "I'll get Slim to open the side stairs, so we can move people faster."

"Glass ceiling." The Parthenon had glass panels to let natural light reach Athena. Gail slapped her thigh and her eDawg moved into heel position, extending his stability handle. Outside, she could hear the rising confusion of the tourists. "There's a school tour in the building. Herd people in here. I'll direct them to the back gallery. No windows."

"Good call. And wrap that art up. We're insured, but I don't want someone to claim negligence."

"Of course." He could not let people just do their jobs. Gail ground her teeth and slid the art on the table into the travel case it had come in. In her HUD, the tornado hit the far end of Centennial Park.

He pulled the door open and propped it in place. "Friends and guests! Please come this way for shelter."

The first person through the door was a woman with two children in tow. Gail set the art back on the table and gestured to the back of the gallery. "This way." She tried to take a step to lead the woman but her right foot stuck to the floor. She shuffled for a moment, desperate to move.

Her AI flashed a message from the medication pump in her side. "Rescue dose?"

Grimacing, she accepted the change with a silent command. Outwardly, she pointed for the woman. "Through that doorway."

Nodding, the woman hauled her children along like bags of wheat. Tourists flooded through the gallery door after her, and Gail stopped trying to move. Some she recognized as members, who came to gallery openings or sat in the calm quiet space upstairs.

Nothing today was calm. Through the open doors at the end of the tunnel leading outside, she could hear a train roaring toward them. The light was the green-yellow of an old bruise and made the people sprinting toward her look ill. Slim staffed the outer door, waving people toward them.

Teenagers on their hoverboards. Parents with jogging strollers filled with toddlers. People clutching their robotic service animals in one hand staggered toward the building. One broad-shouldered pair carried a woman slung between their arms, her robot cat flanking them with stability handle extended as if trying to help.

"This way!" Gail waved all of them past her, to the rear of the gallery. "Meter-thick walls! Solid concrete. Y'all will be safe. It's like a bunker in here."

Aside from the glass panels in the ceiling, of course. But that was upstairs and they were down here.

The train got louder and on her HUD it was nearly on top of them. Keisha, the boy from outside, sprinted in. His violet locs were storm-tossed as he careened through the open door. A pair of women separated from the crowd and ran to him. One woman carried a little girl; the other wrapped Keisha in her arms.

"Keisha. This way!" Gail called to the boy, and gestured for him and his family to head into the gallery, away from the doors.

As they hurried past her, the wind roared down the tunnel, gusting around the gallery with dirt and leaves mixed into the air. Slim was braced against the bronze door at the end of the tunnel. Their feet slid on the smooth floor as they tried to move the door against the oncoming gale.

The door had been designed to shut, but had been decorative for years. The active doors were glass and would do nothing to protect anyone.

Their feet moved and slipped and did nothing. Gail squinted against the grit in the air. A visible wall of churning air filled the end of the tunnel. From the side stairs, Barnum vaulted down, running toward the gallery. Gail pointed. "Door!"

He skidded and changed direction, then pelted down the hall and threw his shoulder against the door next to Slim. Adrenalin seemed to fuel their collective strength. They forced it shut. Slim dropped to their knees and drove the giant old security bolt into its socket on the floor.

With the four-inch-thick bronze doors locked in place, the room seemed almost quiet. The relative silence gave Gail space to hear the people around her. Ragged breaths as if they were still running. Someone was trying not to cry. More than one person was muttering to their phone.

Gail turned from the outer door to face the room full of people. Her feet moved smoothly, as if she were just an old lady with a robot dog. "All right everyone. We're okay. I know that was frightening, but the building is over a hundred and—"

The room plunged into darkness. Upstairs, glass shattered. More than one person screamed, their voices sucked into the wind that kicked around them. Gail froze, but it had nothing to do with Parkinson's. Her heart felt as if it were trying to squeeze out through her pores. The map of the tornado was frozen with it just over the Parthenon and in the corner of the image, an off-line icon flashed.

She swallowed, reaching for her eDawg's handle, and he was right where he was supposed to be. "Wilbur, light, please."

Light glowed from his eyes and in violet racing stripes along his sides and legs. It gave the room the eerie look of a nightclub suspended between songs.

A moment later, an older voice matched her own. "Bilbo, light please." Across the room, another robot dog glowed into shimmering blue existence.

"Isis, light please." A golden emu emerged from the dark.

One after another, service robots lit the dark with a full spectrum. The people that the lights illuminated huddled with their hands laced over the backs of their necks or with a hoodie drawn protectively over their face and eyes. Some of them rocked in place. Others held perfectly still as if they might attract the notice of the storm overhead.

How many people had not made it to the Parthenon?

Gail shook herself away from that thought. A child was crying in the back room, and it sounded as if more than one adult was too. Even with the door shut, the roaring above them made it seem as if the tornado were a monster trying to reach down into the belly of the Parthenon. Gusts buffeted the room.

Slim and Barnum leaned against the walls in the tunnel. Barnum was bent double as if he might throw up. But they were both safe. Although Gail would feel better if they were farther from the doors.

She walked a little ways toward them with Wilbur keeping pace. "Come up here and rest while we wait this out."

Barnum nodded and straightened. "We have water in the gift shop. I'll—"

The doors thudded as if a giant had slammed a fist against them. Gail jumped, letting out a squeak that would have embarrassed her, if everyone else in the room hadn't made a similar sound.

The doors shuddered, rattling against the brass safety bolt. The pair in the tunnel turned and ran as if they had the same mind. Slim vaulted the stairs, reaching back to steady Barnum as he slipped on the top step. Dust-smeared and panting, they sprinted into the gallery.

Barnum stopped by Gail, glancing at the ceiling. So far, none of the thumps had come from above, only a terrifying roaring. "It can't stay over us that much longer."

In the back room, a voice rose from the crowd. "It's okay, love. Shh. Hey. Hey, it's okay. You're safe."

Without thought, Gail moved. She grabbed Wilbur's handle and walked to the back gallery. An adult had a fan in their hands and was waving it in the face of someone who had gone whey-faced with fear. Sweat beaded on their forehead and they were panting. That was a panic attack if she had ever seen one.

The person with the fan glanced around as Wilbur's blue light entered the back gallery. "He'll be okay. I'm his wife—we're fine."

Gail nodded with a smile as if this were part of her normal day. "The tornado is passing, so we should be able to go out soon." She turned to Barnum, who had followed her. "Would you get the case? It's by the wall."

"You didn't put it somewhere safe?" He shook his head. "I told you to—"

"Barnum. We'll talk about this later." She was old enough to be his literal mother. Gail had met her and they were, in fact, the same age. "Would you be a dear and get it for me?"

He huffed, but went to fetch it.

"Really, we're fine." The woman's smile was strained, and she kept fanning her husband who looked as if he would be happier if he actually passed out.

Gail smiled again. "I know you are." She turned and found Keisha, who stood on one foot looking as if he wanted to be useful. "I just promised Keisha here that I'd show him the art for the next installation. Isn't that right, Keisha?"

The teenager's eyes widened for a moment and then he nodded. "Right. Art. I absolutely adore art."

Barnum walked back to Gail and handed her the case. "We are absolutely going to have a talk about this."

"I look forward to it." She gave him her best Bless Your Heart smile and turned back to the room. The air still gusted, but it wasn't as forceful as it had been a moment ago. "Now . . . while we bide our time. My name is Gail Krishnasami, she/her, and I'm the curator here at the Parthenon art gallery. Would anyone else like to see the art for the next exhibit here at the Parthenon?"

Slim raised their hand. "I would."

The golden emu's owner raised their hand. "Me too."

Someone appeared next to Gail and unfolded a chair for her. Her foot stuck as she tried to step in front of the chair and she teetered for a moment, clenching her eDawg's support handle as her right heel raised and lowered in a stutter. She could feel everyone staring at her. A moment later, it unstuck and she dropped into the chair.

Drawing the case onto her lap, Gail peered into the room. "The art that I'm about to show you is eight hundred years old. These are not prints, but the originals. The Parthenon is the home to a recreation of the Pallas Athena that stood in the original Parthenon. She was built back in the 1980s and stands twelve and a half meters high. Who here got upstairs to see her?"

Over half the people raised their hands. The only ones who didn't were the people who had run in from the park. Keisha peered upward as if he could see the massive statue.

"Athena is the Greco-Roman goddess of wisdom." Gail pulled out the tiny portrait in its protective sleeve. "This is Anahita. She's the Persian goddess of wisdom. She's tiny, right? She was painted in the fourteenth century during the Timurid era."

Overhead, winds roared, but below people leaned toward her, squinting through the rainbow of artificial light at the miniature in Gail's hand.

"Now, I'm going to send her around." She ignored Barnum's sudden intake of breath and handed the painting to Keisha. "Keisha will hold her up so you can see. She's painted with egg tempera using single hair brushes."

The teen studied the painting and then very slowly walked down the room, pausing so that people could look at the painting. He waited for people to nod before he moved on.

As he did, Gail pulled another painting from the case and handed it to another person. "This is a painting of dancing dervishes from around 1480 or 1490. It's from the school of Kamāl ud-Dīn Behzād who headed the royal atelier. Slim? Would you take them around?"

They nodded and took the painting, holding the protective sleeve at arm's length so people could see it.

"Now, dervishes were holy people. They tried to approach God by virtue and individual experience. You've probably heard of 'whirling dervishes,' but that was really only something they did during spiritual ceremonies. Other times, depending on region, they were fishermen or mendicants. You can see in this painting, even at the tiny scale, how some of them have been overcome by religious ecstasy."

She pulled out another painting.

Barnum muttered next to her. "Are you going to let people handle every painting?"

Through teeth clenched in a smile, she murmured. "If it helps them stay calm, yes."

From deep in the room, someone said, "Why are they so small?"

Beside her, Barnum made a satisfied snort, and she was honestly surprised that he didn't say, "I told you so."

"Miniatures were intended to be viewed by individuals as part of private contemplation. The idea was that you could keep them in a small book, bound with prayer or poetry, to have readily available in a pocket."

Keisha looked down the room. "Like watching a movie on a smartphone?"

"Y—yes." Gail nodded. "Or seeing it on a HUD. When you are close, it can appear to take up your whole field of vision."

"I can't get over the detail."

"Did you see the golden flowers?"

"Look at the horses!"

The conversation rose and twirled around them like wind. People leaned together, staring at the tiny paintings as they paused in front of them. While they waited for the "all clear" signal to sound, Gail showed all fourteen miniatures that she had prepared for the collection. She answered questions.

At some point, her dyskinesia kicked in, making her right shoulder move in annoying rhythmic pulses. By that point, the paintings were all out of the case, being carried by volunteers in a mobile living gallery. Behzad's *Advice of the Ascetic* had made everyone gasp in wonder at the heavily illuminated border, with its dense field of gold leaf animals cavorting on a deep blue background.

She'd asked Barnum to carry that one.

The man who had the panic attack had stared at a scene from Attar's *Conference of the Birds* as if the hill and the brook running down it were a lifeline. The person carrying that one had paused in front of him and waited as patiently as a statue.

Slim carried the painting of Anahita back to Gail and paused, looking out the gallery doors. "Is it... It's quiet."

Gail closed her eyes and all she could hear were conversations about art.

"The egret has a fish!"

"Look at the expression on her face."

"So this one was from the Tabriz style?"

The roaring train had faded. Only a faint breeze stirred the air. Gail opened her eyes and looked for Barnum. He was holding the Behzad carefully. "Right, so during this period they prized heavily decorated borders."

Keeping her voice low, Gail asked Slim, "Do you want to go up and check?"

They handed her the Anahita. "I'll be right back."

Gail held the painting in the blue light from Wilbur. It would have been nice if she had somehow seen a hidden feature that could only be viewed in the precise spectrum her robot dog emitted. But the reality was that the single blue source washed out the painting so it was nearly grayscale. Even so, the precision of the lines in the tiny painting was captivating.

Slim clattered down the stairs outside the gallery. "The sun is out! It's passed!"

A moment of shocked silence stopped all conversation, and then a cheer went up. Gail sagged against the chair back with relief. The tornado had probably passed them before she'd gotten the first painting out of the case.

She slid the Anahita back into the case as people stood, stretching from the floor. Barnum hastily handed her the Behzad, and jogged across the gallery to the tunnel. People stirred and began to follow him, eager to get out of the building.

The next paintings were harder to get back into the case, as her body decided to move randomly with dyskinesia. Gail gritted her teeth, trying to slow down enough to have control as she put each protective sleeve back in place. By the time she had retrieved the last one and stowed it in the box, she became aware of a general restless murmuring.

Frowning, Gail patted her left thigh. "Wilbur, heel."

He was at her side before she finished the final L in the command. Gail stood, peering past the people clogging the exit from the gallery. From the front of the group, someone said, "Well, damn."

"What?" She stood on her toes, trying to see past the crowd.

"Door won't open."

Whatever had thumped against it had warped it so it wouldn't open. Setting the case against the wall, Gail chewed her lower lip. There were two additional exits upstairs. Two massive sets of bronzed doors that had been shut for as long as the ones down here had been open.

Before Barnum had started working here at any rate. Gail worked her way through the crowd. Past them, she had a good view of Barnum and Slim trying to force the brass security bolts up from the ground.

"I'm going to try the upstairs doors."

A guest turned, startled. "Those open?"

"It was the main entrance until the 1980s." Gail walked toward the stairs, her feet moving more smoothly now that the rescue dose was thoroughly in her system. Her shoulder jerked in syncopated time, but she didn't need her shoulder to climb stairs.

Oddly, stairs were always the easy part for her. She and Wilbur walked up the stairs into daylight. Most of the roof was gone. Water covered the polished concrete floor of the Naos. Gail kept her gaze down, biting her lip, and did not want to look at the dais at the far end. All she really needed was to find out if the doors worked. And yet...

Athena was still there.

Gail's knees went weak. Water dripped down the gilded skirts of the massive stature. Leaves and paper were plastered against her, but she stood.

"Whoa." Keisha had followed Gail up the stairs. "Whoa."

He was not the only one, either. Gail had acquired a retinue of visitors to the Naos. The mother with the children. The woman and her husband, no longer panicking. A dozen other people that she hadn't spoken with downstairs.

She gestured to the giant statue. "For scale, the figure of Nike in her palm is a little over two meters tall."

"Whoa." Keisha said again.

Gail smiled at the teen. "I did tell you that she was worth seeing."

Even with dirt coating her gilding—or maybe because the ceiling was open above her, maybe because Athena had come through the tornado miraculously unscathed, she seemed more radiant than ever. Although, considering that she had rebar going through the Parthenon and into the bedrock below, maybe it was less miracle and more engineering.

Keisha looked back toward the stairs. "Do you think we could bring Anahita up into the light?"

"Not with the water." Gail shook her head and walked with Wilbur to the massive doors. "But these are the largest matched bronze doors in the world, if you're interested."

"Whoa."

Gail slid a panel aside in the door to reveal the small handhold hidden behind a lion's face. Gripping it, she disengaged the lock and pulled back on the door.

For a moment, she thought it wouldn't move either, but the door slowly opened on the brass support arc embedded in the floor. As it swung open, her HUD suddenly pinged.

*Connection!*

Messages downloaded in a storm of alerts and pings. Behind her, the conversations died away again as other people got the same sudden influx of communication.

Bobbi had left two full screens worth of messages. Her wife's texts got shorter and more frantic as—

"Gail!" Barnum jogged up to her, tilting his head back to stare at the massive opened door. "Wow ... I don't think I've actually seen this open before. Good thought."

"Any luck downstairs?"

He shook his head, turning to face back into the Naos. "All right, everyone! We've got the exit open here. If I could have some volunteers to help the people with strollers or mobility challenges come up the stairs, I would greatly appreciate it."

People applauded.

Gail clenched the support handle on her eDawg so hard her knuckles turned white. It was fine. People were getting to leave and they were safe. Her shoulder twitched faster and harder. The onboard AI sent a notice about adjusting the dose of Cyphrenine to control the dyskinesia. It would leave her nauseous and constipated later, but she OK'd it because anything was better than twitching.

As people filed out though, they slowed, looking at Gail. "Thank you for the art history!"

"You were wonderful downstairs."

"I can't wait to come back when the exhibit is open!"

"Anahita and Athena for the win!"

"GAIL!" From behind her, Bobbi's voice could have cut through concrete.

Gail turned so quickly that only her eDawg kept her from losing her balance. She gripped his handle as he moved with her, countering her weight until she had her feet under her again. Barnum got a hand under her elbow and steadied her. She flashed him a smile of thanks. "Excuse me. My wife—"

"Go!"

With her robot dog at her side, Gail walked out of the Parthenon into the cool, damp aftermath of the tornado. The great green lawn had a swath of brown dirt plowed through it. A massive tree trunk filled the tunnel to the basement. On the steps and colonnades of the Parthenon, branches and debris clogged the spaces between the columns.

In the middle of that, her wife was climbing the steps, looking frantically at her phone. "GAIL!"

"HERE! Bobbi, I'm here." She stepped past part of a park bench and used Wilbur to get down the stairs. "Bobbi—"

Her wife sprinted up the stairs and wrapped her in an embrace, sobbing. Until that moment, Gail had not let the terror hit her. She dropped her eDawg's support handle and held her wife. It was hard to say which of them was supporting the other.

"Are you okay?" Bobbi stepped back, sliding her hands up to hold Gail's shoulders. "I've been so . . . I couldn't tell where you were."

Wincing, Gail nodded. "I'm fine. I'm sorry." She gestured back to the Parthenon. "It's a bunker. We were all fine, but without power. Listen. Earlier . . . I'm sorry I turned off my tracker."

"I shouldn't have pushed." Bobbi's long silver braid had strands of hair escaping the plait as if to echo her agitation. "But you're okay?"

"Totally fine." She almost suggested that they head home, but the art was still out in the gallery. Even though the art was in its case, she didn't feel good about leaving it there. "I just need to . . . I'm sorry. But I need to finish up some things here."

She could see the protest building in Bobbi's eyes like a storm. Her wife opened her mouth and then bent her head. "Of course." When she looked up, the storm had passed. "Can I help?"

Gail reached for her wife's hand and squeezed. "Thank you."

When they turned and started back up the stairs, Barnum was standing by the side of the massive doors, staring at Athena. He jumped a little as they reached him. "Oh, hey . . ." His eyes were red at the edges. "About the exhibit. I was wrong. Do what you wanted to do, okay?"

Because she was a grown-up, she did not do a victory dance or make a snide comment or act out any of the various forms of *I told you so* that filled her brain. She looked at Athena and tried to follow her example. Gail smiled at her boss. "Thank you."

Bobbi raised her eyebrows after they were past. "What was that?"

"Victory through wisdom." With her wife on one side and her support robot on the other, Gail went into the basement of the Parthenon and did her job.

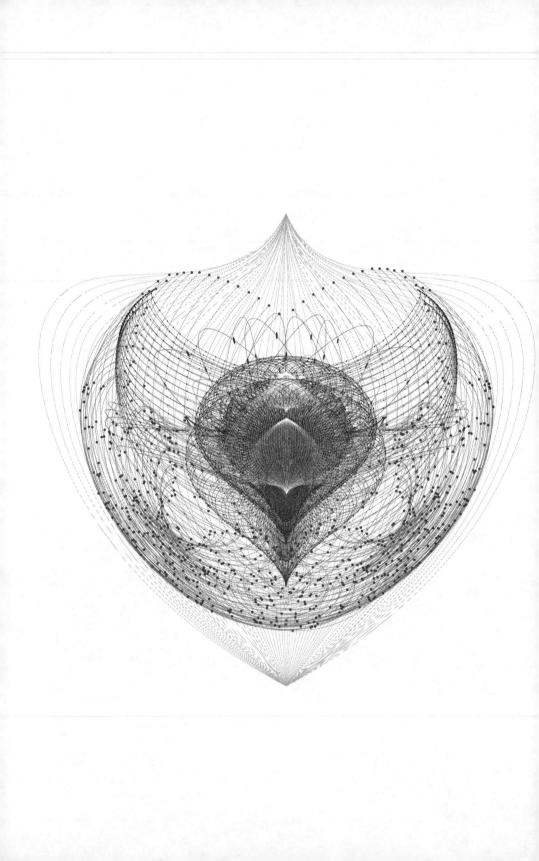

# 6 YOUR BOYFRIEND EXPERIENCE

## James Patrick Kelly

**"IT'S NOT A DATE," JIN SAID.**

I couldn't believe we were having this conversation. "Then tell me again what it is." We'd been snuggling as we played our therapy adventure, but now I scooted away from him on our couch.

"Like I said, just a field test of Partner Tate." He leaned forward and scooped up a handful of wasabi popcorn from the bowl on the coffee table. "We want to see how our new partner does in real-world encounter situations." On our living room screen, Jin's Tik-Toc avatar swiped a sword from the rack on the castle wall and tossed it to my gingerbread man. "It's a *simulated* date, Dak," he said, intent on the game. "And a chance to see what I've done on this project."

I pointed a pause at the screen. "And you're asking me to encounter your playbot?"

He stiffened at the interruption, but to hell with him. The point of this stupid couples game he'd brought home was to foster teamwork and build trust, but my boyfriend hadn't been playing fair in the real world for weeks now. "And just where will this not-date take place?" I asked.

"You could go to one of those fancy restaurants you're always talking about. Stage Left, or the Ninety-Eight. On Motorman's tab." His hand twitched, but he knew better than to restart the game. "Say a club afterward."

"A club? What club?"

He glanced over at me and saw trouble. "Or a puzzle palace, bowling, whatever you want."

"Oh, perfect. Maybe we'll run into your mother at her league."

"Look, Dak, I love you. You're my ..."

"... partner." I hated it when he said that word just to keep the peace. "I'm your partner, your boyfriend experience. Like Partner Tate."

His lips parted as if to reply, but he thought better of it. He covered his indecision by reaching for more popcorn. His tongue flicked a single kernel into his mouth.

"And after some puzzles, then what?" I'd never liked the way he ate popcorn. "Back to his place?" Jin was patience—nibbling one kernel at a time—and I was impulse—chomping my snacks by the fistful.

"He doesn't have a place," Jin said. "He's a prototype, lives at the facility."

Why was I so upset? Because I couldn't remember the last time Jin and I had been on a date. How was I supposed to get through to this screen-blind wally who had the charisma of a potato and the imagination of a hammer, and who hadn't said word one about the Shanghai soup dumplings with a tabiche pepper infusion that I'd spent the afternoon making?

"Just because we call them partners doesn't mean you have sex with them," he said, missing the point. "If you don't want to have sex with Tate, it will never come up. He doesn't care."

I wanted to knock the popcorn out of his hand. Instead I said, "Okay." I flicked the game back on. "Fine." I huddled on far side of the couch. "You win."

"Thanks, Dak. I'll set it up." He turned back to the screen. "Oh, and Aeri wants to meet you, if that's okay."

"Wonderful," I said, without enthusiasm. Of course, Jin was oblivious. I wasn't sure which was worse: meeting Jin's new playbot or Aeri Dashima.

ALTHOUGH MOTORMAN HAD JUST A FRACTION OF THE HUNDRED BILLION DOLLAR playbot market, the company had made more than enough to pay for the lavish headquarters where I got my first look at Partner Tate. Of course, most playbots—don't call them sexbots!—had female chassis, since the breakthrough customer base had been straight men. By far the most successful playbot company was Zfriendz, which controlled almost 60 percent of the market; they sold five playgirls for every playboy in their showrooms. But the popularity of the sexualized Girlfriendz® lines meant little in the more challenging, but still lucrative playboy market, leaving room for niche companies like Motorman to compete for the business of women and gay men.

And so here we were: Jin, project manager of the Partner Tate team, and Aeri Dashima, CTO and co-founder of Motorman. She eyed me across the conference table, a disconcertingly short woman with silvered hair, pearly skin, and eyes bright as brushed steel. She wore a DeGoss shapesuit that cost as much as the downpayment for our condo. I might have guessed that she was in her late forties, even allowing for stemcare and her state-of-the-art body conditioning armor, but Jin had told me that she'd started her first company with her late husband some eighty years ago. I could see that now; she had a young face but old hands.

Aeri was so petite that she struck me as birdlike. Her head cocked to one side, she seemed amused as I pressed my thumb to screen after screen that Motorman's lawyer brought up on her tablet.

"Sign here." When the lawyer saved and flicked a form away, another replaced it. "And here's your nondisclosure. Again, you're welcome to read. Feel free to take all the time you want."

"Why?" I said. "Is any of this negotiable?"

The lawyer tweaked regret into her smile. "I'm afraid not."

Aeri might have been the only one in the room who was comfortable at this meeting. Every so often I'd steal a glance at Jin to see if he was sufficiently grateful that I was doing this for him. But he wasn't looking at me, the tablet, the lawyer, or his boss. Instead he chewed his lip and peered out the window as if plotting his escape: crash through the floor-to-ceiling windows of the conference room, swim the koi pond, sprint through the Japanese garden across the parking lot and into the piney woods that surrounded the Motorman headquarters.

"Do I remember that you're a chef, Dakarai?" Aeri asked.

"I collect cookbooks," I said. "And I curate a specialty cuisine forum."

"He has three thousand supporters," said Jin. "Dak is a fantastic cook. Neo-infusion."

"Is that the one where you sprinkle gold onto everything?"

"Gold is inert," I said. "Indigestible. You might as well suck on a nail."

"Liability waiver," the lawyer murmured. "Here and here."

Aeri shifted in her chair. "So neo-infusion is...?"

Did I want to explain myself to this woman? "So once upon a time infusion was mostly about infusing dishes with cannabinoids and other terpenes. Neo-infusion is more about borrowing flavors from other plants and mixing them across cuisines. And it's not just infusion." I made eye contact to show Aeri I wasn't intimidated, but then the lawyer guided my hand back to the screen. "I use decoction and percolation. Tinctures."

"Making alcohol the solvent instead of water," said Jin.

"I know what a tincture is," said Aeri. "Chemistry 101. But terpenes?"

"Those are just fragrances," I said. "Think essential oils, like they use in perfumes and aromatherapy."

"Surveillance consent," said the lawyer. "Here, please. And last, the insurance acknowledgment." When the tablet had recorded my last thumb, she waved to save all. "I'm sorry," she said, "but does Partner Tate eat?" She leaned over to retrieve her briefcase from the floor. "Feeding him might have some liability implications."

"He does and he doesn't," said Jin. "He consumes samples, but doesn't need food. We've designed him to be able to taste what his human companions are tasting."

"My goal for the Partner Tate line is that he share in his primary's enjoyment as much as possible." Not only did Aeri own a controlling interest of Motorman, but she remained its star personality engineer. "But I think we've had enough law for now, Devya. Send us a memo." She flicked a finger at the lawyer as if she were an app to be dismissed. "Thanks so much."

The lawyer left, and Jin twitched back into the room from wherever he'd been daydreaming. "Ready for the introduction?"

When Jin left, it was just the two of us: the man and the woman in his life.

"We appreciate how understanding you've been about Jin's workload," Aeri's mathematically thin eyebrows made her gaze seem more intense. "He has accomplished a great deal. It's an important project, both for his career and our company."

More important than our relationship? My stomach churned; had Jin shared our problems with this AI plutocrat?

"You are nervous, Dakarai."

"No," I lied. "Should I be?"

"Not necessarily." When she tucked her legs under her on the stool she seemed to rise above the conference table. "But I believe you are. I have a second sense for these things."

"I suppose people will stare," I said.

"They will, some more than others."

She was right about my anxiety, only I was more worried about my future with Jin than any encounter with her prize playbot.

"This isn't our first field test," she said. "Of course, we're interested in how Tate interacts with you, a civilian. But we're also interested in how onlookers react."

"Onlookers?" I grimaced. "Sounds more like an accident than a date."

"We know that some will be distressed by Tate. They'll tolerate play-bots in bedrooms, but not out in public."

"Prudes and holy joes and never-bots. But now that Jin's got your battery problem solved, the Tates won't have to hang around their rechargers."

"There's still work to do on the power problem. An advanced model like Tate draws 300 watts an hour, a serious burden even for fluoride ion cells. You'll have a five-hour window before Tate's batteries discharge."

"Right, Jin explained the curfew." I drummed a finger on the conference table, counting the hours: one, two, three, four, five. "Midnight, or my date turns into a pumpkin."

"At least he won't be wearing glass loafers." Jin entered, grinning like an idiot, then stepped to one side and gave us a proud flourish.

He'd never shown me pictures of Partner Tate. Now I knew why. Yes, I was surprised. Not shocked, *surprised*. But it made sense in a twisted way. Jin beamed as he waited for my reaction. I caught an image of my boyfriend as the nerdy kid he must have been, standing beside his first-place project at the science fair. I swiveled in my seat to check Aeri. She was amused, seeing as the joke was on me. With all three of them staring, I decided it was easiest to meet Partner Tate's impassive and all-too-familiar gaze.

He wasn't me, exactly, but we could've been brothers. We were the same height, but a hundred and seventy centimeters was average, as Jin always teased when I complained about being shorter than him. Tate and I both looked discreetly fit, but not ripped. Straight black hair, brown eyes—his ears were flatter. Neither of us was handsome. Fine-featured; that's what my mom used to say. He was dressed better than me, in a tailored blue suit and high-collared silk shirt. The loafers appeared to be real leather.

"You must be Dakarai," said the playbot. I thought his smile needed work. "I'm Tate." He stepped around the table and offered me his hand.

I gave him a bleak hello. His shake was convincing: two pumps and a release in the proper five seconds. But was the palm too dry? The grip weak?

I guessed I should pretend he was real, even though this was where he'd been made and these were the people who'd made him. "I'm struck by the likeness, Tate." But I wanted to test the rules. "I wonder how anyone is going to tell us apart?" I shot Jin a glance. "Or is that the point?"

"Perhaps our family resemblance might serve as protective coloration." Without asking, Tate took the seat beside me and nodded a greeting at Aeri across the table. "Jin and I hoped you'd be pleased. But perhaps my looks make you uncomfortable, Dakarai? We've discussed making adjustments."

"Of course, the production models will be totally customizable." Jin settled onto the edge of his seat. "But Tate is one of a kind."

Tate smiled and nodded. "Thank you, Jin."

I wondered what kind of relationship Jin and his playbot had. "Not uncomfortable, no," I said. "You know, it's been years since I've seen a playbot up close, but Partner Tate here is a shock. He's jumped clear over the uncanny valley. That's going to take some getting used to."

"Call me Tate, Dakarai." Tate patted my hand—just a feather touch. "And don't worry, I've still got plenty of tells to give me away, once you know what to look for."

"We've got Tate on maximum simulation," said Aeri. "You can dial him back to be more robotic if you like."

"Why?" I asked. "Is that a thing?"

"Touch him," said Jin. For a moment I thought he was talking to Tate. "Go ahead. Try his cheek."

The playbot nodded an encouragement and leaned forward. This was creepy—something no real person would do. But it was their show. I pushed at the side of his face my forefinger. His understructure was a bit too rigid, even when he opened his mouth. I let my finger slide down his cheek. Although I couldn't see it, I could feel the sandpapery hint of a stubble. Sometimes I needed to shave twice a day. One of the things I loved about Jin was the silk of his jaw; he'd never had luck growing a beard.

As I let my hand fall, Tate made a quick feint as if to bite it. His teeth clicked on air. Then he gave me a wicked chuckle.

It took me a beat, but I pushed a laugh out, too.

"Don't worry, Dakarai," the playbot said, "I promise to respect Asimov's Three Laws."

"Asimov?" I frowned. "I don't know who that is."


WHEN JIN CAME HOME FROM WORK THE NIGHT BEFORE MY DATE, HE WAS EVEN MORE jangled than usual, so I sent him to meditate while I finished making dinner. To help smooth out our wrinkles, I improved the sugar syrup I'd infused with vanilla, ginger, and rosemary by adding a hefty dose of golden dragon cannabis tincture. I drizzled this over a fruit salad.

Half an hour later, Jin rose from his yoga mat, came up behind me and caught me up in a fierce embrace. "Smells great, babe," he said, nuzzling my neck. "I'm hungry."

"Good to know," I said. "But I can't finish spicing the stew if you're pinning my arms."

He released me with a kiss. With a twinge of desire and exasperation, I wondered what this charm offensive was about. The timing felt off, and Jin had never been much of a hugger. He preferred to fondle, or maybe brush a hand down my jaw. I tipped the mixture of toasted coconut, turmeric, Makrut lime leaf, and asam keping into the simmering chicken redang. "Five minutes," I said. "You can take the fruit salad and set the table."

"Wine tonight?"

"Not for me." I nodded at the bottle of golden dragon on the counter.

"Ah," he said. "Excellent choice."

We made short work of the fruit salad; the star fruit was a little too apple-y, but the cantaloupe was sweet and musky. Over the two years of our relationship, I'd persuaded Jin to pause between courses, since I practiced mindful eating. But that meant we had to talk while we digested and I wasn't sure what to say to him.

"I had another meeting with Aeri today." Jin poured us iced mint water. "She likes you."

"Could have fooled me."

"And Tate liked you, too."

"Like he had a choice."

"Of course he does. I told you, that's part of the design's verisimilitude." We were sitting opposite each other. He shifted the stem vase with its single carnation to one side so he had a clear view of me. "Come on, boyfriend, pull up," he said. "You're losing altitude."

"Sorry," I said. "But you never said anything about the way he looks. And now I'm wondering why not."

Jin sighed.

"A hell of a surprise." I tried to grin. "I suppose I should be proud."

"But you're not."

"I'm still processing." I held up a hand. "But I don't hate it."

"Good, because I like his look." He gave me his best leer. "Reminds me of a certain sexy someone." He reached into his pocket and took out his phone. "But maybe we shouldn't have surprised you. That was Aeri's idea. The surprise, I mean."

"Why should she care? She doesn't know me."

"She wants to." He tapped the screen. "But she likes to be in control, Dak. You probably saw that. This is her pet project and I'm her pet designer." I tried not to be annoyed as he flicked through messages. He knew my dining policy: phones and food don't mix. "But I can handle her as long as I have you with me."

I wanted to believe that. "Okay."

"She presented me with this." He passed the phone. "A bonus for my year of Partner Tate."

Was the giddiness I felt coming from the golden dragon or all the zeroes? Jin had always made way more money than me; my income from the forum paid for my kitchen equipment and our groceries and not much else. But the amount on the screen represented almost a year's rent for our apartment. Or a vacation for two on the moon. Or one of Motorman Corporation's top-of-the-line playbots.

I set the phone screen facedown on the table. "Wow, Jin." Then I flipped it over again to see if I'd read the numbers right. "Just wow."

"I know." He was nodding. "And a promotion, too. Director of Partner Development." He beamed the same way as he had when he'd introduced Tate. But my pleasure at our good fortune turned when I realized what this meant. I'd heard horror stories of how the stars in Aeri's inner circle slept under their desks or on daybeds in Motorman's playroom. There would be all-day meetings and midnight phone calls. Was this the down payment for what remained of Jin's scarce free time?

"So another project coming up?" I said. "Son of Tate?"

"Sure." Jin came around the table to me. "Always something new." I thought he meant to clear my empty bowl so I stood to fetch the chicken redang and coconut rice. But he put hands on my shoulders to sit me back down.

"I've been so worried that I'm losing you." He brushed the back of his hand against my temple. Then he dropped to his knees.

"Jin, what the fuck?"

"I've been busy and I haven't had time for you. For us." He took one of my hands. "I know I live in my head too much and keep forgetting to compliment you for all the things you do. I snore and I'm finicky and I can't tell a joke. Meanwhile you're smart and sensitive and creative and a great cook and a star in bed and you're the most important thing in my life and I ..." He fumbled a box out of his pocket and thrust it at me.

His face was as big as the moon. How many feelings can a guy have at once? Did I believe him? Did I want him? Why did this have to happen when I was high? I opened the box with a thrill of joy and dread. Inside was a retro house key that appeared to be made of gold. It was an uncut blank: no notches or ridges.

Now I was officially confused.

"The key to the house we're buying together," he said. "That's what the bonus is for. Aeri's wedding present. Dakarai Delany, will you marry me?"

I didn't have an answer. I needed to think, but all my thoughts had turned to steam. I couldn't leave him on his knees, so I pulled him to his feet. We stared at each other and his expression froze.

"Jin . . ." I said.

"Oh god, I've fucked up again." He took a step back. "I'm no good at this." And another. "I'm better with bots than I am with my own boyfriend."

"No, Jin. It's fine." I thought he was about to run out of the room. "Better than fine. I'm just surprised." I took one of his hands. "Look, I would very much like to marry you. But who will I be marrying? Will it be workaholic Jin or the Jin I fell in love with?"

"The Jin you fell in love with was a pretty hard worker."

"Yes he was." I smiled. "Work is okay, but too much is too much."

"Do I have to leave Motorman? Is that it? If I have to, I will. Absolutely I will."

I loved this man and I knew everything depended on my saying the right thing. Whatever that was. "I don't know if you have to." My throat closed to the size of a straw. "Do you think you do?"

"If I answer that question, will you answer my question?"

"Fair enough. All things being equal, which they aren't because you're the best, yes. I will." I kissed him. "Marry you. And your question?"

"I think . . ." A shadow passed over his face. "Yes, maybe I do need to quit."

Not exactly what I wanted to hear. Did I really want to make him give up his career at Motorman?

"Let's talk about that later." I caught him in an embrace. "Right now we need to celebrate. Come to bed with me, you crazy, lovely boy. The stew can wait, but I can't."

I DIDN'T KNOW WHAT TO EXPECT FROM MY EVENING WITH TATE, SINCE MY DATING resume would have fit on a lottery ticket. I'd met Jonoh, my last boyfriend, in college and we'd lived together on and off for eight years. A couple of times I'd left, but I always came back to him. The one time he walked out, he kept going. I was teaching social studies in middle school then. At the end of the school year, I realized that I was done, gave my notice on an impulse, and worked retail for a couple of years. I was the sorrowfully celibate assistant manager at the Cook With Us franchise at the White Rose

Mall and had just started my forum when I met Jin. He'd come into the store shopping for measuring cups.

So I'd been out of the game for a while. But then again how many dates had Tate been on?

"Four," Tate said. "One each with Aeri, Jin, Gunter Kruger, and now you."

We were strolling down Washington Street toward Union Tower. "Did you enjoy them?"

"Sure. But I'm still working on my style."

I wanted to ask about his date with Jin, but decided to work up to it slowly. "So, you went out with Aeri Dashima. Were you nervous? I mean, is there even an algorithm for that?" I felt like I was nattering.

"Software for all occasions. Aeri likes to say that her bots are nothing but windup toys; the algorithms and memory are who we are." He gave me an ironic salute. "Why, are you nervous, Dakarai?"

"Nervous?" I thought about it. "I was."

"Was. But not now?"

The light changed and we stepped into the crosswalk. I'd left Jin at home for the evening so I didn't have to pretend for his sake. I could say what came to mind. "The kiss of the golden dragon." I reached into my pocket and showed Tate the vial.

"Ah," he said. "Jin mentioned you were a THC fan."

What else had he discussed with his playbot? "We both are."

"Should I get high, too?"

"What?" I was so astonished that I stopped in the middle of the street. "How?"

"Like I said, I'm made of algorithms." He tapped his head. "A switch in the code and I'm your neighbor in Fun Town."

"Okay." We continued walking. "Knock yourself out. That way when I catch myself jabbering I won't feel self-conscious."

**STAGE LEFT WAS ON THE TOP FLOOR OF UNION TOWER WITH A VIEW OF THE RIVER.**
An express elevator opened onto a long interior hallway. On one wall were windows overlooking the busy kitchen stations. The maître d'hotel stood at the far end. We kept him waiting because I held Tate to a slow walk. In the kinds of restaurants I could afford, bots did prep and cleanup, but as I browsed the window wall, I saw only human staff. I watched the line chefs and kitchen hands at work. They had some of the same microplanes,

Jaccards, and frothers I used in my own kitchen. But the cultured meat processors and carb brewing tanks were beyond anything I could afford.

"I don't see Sofia Vasquez," I murmured.

"Who's that?"

"The chef de cuisine. Also the food host on *Stylelife*. Looks like her sous-chef is running the kitchen tonight."

"Is that bad?"

"No, but she's the brand name here." I glanced at him and laughed. "Go ahead and judge," I said. "I'm not ashamed of my celebrity crushes."

"Crushes? How many do you have?"

"Slow down, pal," I said. "It's only the first date." I nodded at the maître d'. "Reservation for Delany."

"Welcome to Stage Left, gentlemen." His eyes flicked from me to Tate. Could he tell that Tate was a playbot? "This way."

My first impression of the dining room was of a vast and empty space filled with pastel light. But it wasn't empty; twenty or so tables were set against the two side walls, little stages lit by an array of spotlights. The back wall was a single window that looked at the glimmering city and its black river. The décor was all sharp edges and hard surfaces: tables of brushed aluminum, chairs made of rectangular slabs of glass. The tables were extravagantly spaced—the restaurant could have accommodated three times as many diners. Those who'd arrived before us were scattered so far apart that we heard none of the usual dining gabble. Or maybe they were hushed out of reverence for the conspicuous privilege on display. At the center of the space was a low dais on which plates of exotic food and glasses of mysterious drink appeared as if beamed in from the twenty-third century.

We were shown to our table. "I've chosen Kylie to be your server tonight," said the maître d', beckoning to the line of waitstaff standing at attention.

"Excuse me," said Tate, "but will the boss be in at some point?"

I grimaced. It was a question I might've asked myself had I been bold enough. But where was the respect? Sofia Vasquez was the chef de cuisine of the finest restaurant in the city.

"It is the chef's custom to arrive around eight." The maître d' made no attempt to hide his disdain for Tate's word choice. "So yes, sir. Momentarily."

Tate grinned at me. "I only ask because Mr. Delany is a fan."

"Ah." The maître d' tugged at his lab coat, although it didn't need adjusting. "As so many are. Enjoy your meal, gentlemen." He returned to his station.

"The *boss*?" I said. "Where did that come from, your drug algorithm?"

He snickered. "That's what she is."

"This isn't some pizza joint." I wagged a finger at him. "It's culinary theater. We all need to act our parts."

"In that case, that guy is a ham. Even worse than last time."

I leaned back in my chair. "You've been here before?"

"With Aeri." He unfolded his napkin.

Before I could digest this, our server arrived, all grins and goodwill in her sky blue lab coat. "Good evening, gentlemen. My name is Kylie and I'll be your guide to the gastronomical arts of Sofia Vasquez."

Tate covered a laugh with a cough.

"Is this your first visit to Stage Left?"

"Yes," I said.

Tate gave her a sly finger wave. "Second."

She seemed taken aback, before apparently recognizing him. "Your meal tonight will be presented in seven acts." She offered us the famous edible menus. They were as advertised: thick and brittle as melba toast. On their smooth upper surfaces, they looked like programs for a play. "You'll have difficult decisions to make on the middle three acts and the finale. May I bring you something from our vapor bar?"

"Can you make an espresso martini?" I asked.

"Absolutely." She beamed approval. "Something for you, sir? Mr. Tate, am I right?"

"Water will be fine."

"An amuse-bouche then, in celebration of your visit." She stepped aside. A follow spotlight tracked a new server, wearing a full Tyvek clean room suit, including face mask and gloves, from the central dais. "A beetroot and horseradish macaron," announced Kylie. "Sofia recommends that you eat them all at once."

The overhead lights washed our table in a ruby glow. The macaron was exquisite. Round and the size of a Ping-Pong ball, it was so light it might have floated off the plate. Deep red aerated beet halves sandwiched a creamy horseradish filling. It was tricky picking it up with a fork but worth the effort. Spicy and sweet with a mustardlike finish. It was heaven's own appetizer.

"Good?" Tate watched for my reaction. "Want mine?" He held up his plate. "It stains my tongue purple."

I didn't want anyone to see a stack of plates in front of me, so I passed him mine. "What did Aeri think of this place?"

"I don't think she cared for the food, but she was impressed by the bill."

I chased Tate's macaron around the plate.

"You know, I tried to get Jin to come here," Tate continued, "but he said he was waiting until he could bring you. We found a Fujian cart in Little Harbor instead and ate in the park."

I set my fork down and the macaron escaped. "So what did you two talk about on your date?"

"You, mostly." Tate broke off a corner of the menu and showed it to me. "I understand this is made from cassava. Printed with soy ink." He popped it into his mouth and crunched. "Tastes like stale potato chips."

"You're not supposed to eat it until after you read it."

"Why bother?" he said. "I'm having what you're having."

"You're enjoying this, aren't you?"

"Your company? Yes." He didn't bother to hide his delight. "Things are going very well, don't you think?"

Maybe it was the golden dragon, but I had to laugh. Of course he was flirting; he was a playbot. "Look, Tate, you can't say you talked about me and then change the subject."

"I realize that, Dakarai." He bowed. "You have only to ask."

"Pardon me, Mr. Delany." Kylie was back already. "Your sniff." She offered me the straw, and Tate settled back in his chair to watch the ritual. The overhead lights turned a misty blue as she poured the martini into the sniffer and picked up a dry ice cube with silver tongs.

When I gave her the sign, she dropped it into my drink. Jin didn't like me to smoke cocktails, because he said it made me silly. But I was with Tate tonight so I poked my straw deep into the roiling cloud of alcohol and sucked up vapor for a good ten count. A magical fog of vodka, vermouth, olive brine, and coffee oils coursed straight from my lungs to my bloodstream and right to my brain. It felt like an ice cream headache, except warm and fizzy.

"Very good," I said. "Thanks."

"You're welcome." Kylie poured water for both of us. "Take as long as you want with the menu," she said, pointedly ignoring the crumbs on Tate's bread plate.

When she was gone, Tate said, "You're ahead of me again. Should I keep up?"

"Keep up?"

"I can match you drink for drink." He tilted his head toward the sniffer. "Only in software." A last wisp of vapor curled in the spotlight. "And Aeri won't have to pay for it."

I grinned. "If we're both tipsy, who'll be the responsible party?"

"Ha!" He lifted his water glass. "Responsibility is overrated!"

Was I charmed? I was. This was like no other date I'd ever been on. Tate was nothing like Jin, and I couldn't help but appreciate the difference. Except hadn't he helped create Tate? Was this playbot my boyfriend's ideal self? "So," I said, "what do you know about Jin and me that I don't know?"

He gave a lifelike twitch, as if in reaction to his virtual shot. "He loves you," he said at last.

"Okay." I forced myself to stay cautious.

"He claims that he tells you a lot. Every day, says he." Tate seemed to be doing a difficult calculation. "But not sure you believe him. Or even hear."

The truth raised a burn of embarrassment on my cheeks. "And?"

"Convinced you'll leave when he screws up."

"That's not true." I blurted this, and then thought better. "I suppose it depends on the level of screwing up."

"Point, yes. He can't tell his best move from his worst. I mean, thinks you want him quitting Motorman, but then will lose the bonus. Which means no Jinny and Dakarai dream house."

"He told you about that?"

"And getting married." Tate was doleful. "Except not, maybe, since you two have to work all kinds of issues."

"Issues." The word seemed to twist in my mouth. "Right."

He was fidgeting. "I love the engagement present, though. That key thing? Not sounding to me like something himself thought of, but maybe he looks to the internet. Shows he's trying, no?"

Tate wasn't slurring his words, but his speech sounded compromised. "You're on his side," I said.

In reply, he pushed his chair back and bent to peer under the table. He stayed down for a good thirty seconds.

"What?" I leaned over too, but saw nothing.

"His side, your side." Tate straightened. "All sides same me." The fingers of his right hand began to drum on the table. "I love Jin and Jin loves

Dakarai. Logic is Tate must love Dakarai." His fingertips hit hard; they sounded like rain spattering on a windshield.

"What are you talking about?" I glanced around to see if anyone had noticed his odd behavior. "Tate, are you okay?"

"I know, I know, I know, I know." He was looking at his hand as if it had wandered over from another table. "I hear myself saying things that aren't true. How love I anybody? No I, here. Not even conscious. Just a mess of algorithms, enhance your boyfriend experience experience. Your next tasteful but tasty sexual encounter, brought you by Motorman. Only no sex with Jin." When he made a keening sound, like servos under extreme load, the man at the next table turned. "Doesn't want to sex with me because. Just because." In the silence, his voice carried and the line of servers rippled.

"Tate," I hissed. I closed my hand over his to stop the drumming. "People are staring. What's wrong with you?"

"Maybe overstimulated. Intimidated. Inundated." His gaze flitted from one of my eyes to the other, as if he were seeing two of me. He whispered to us, "Would you ever sex with me, Dakarai?"

"Enough!" I crushed his hand flat against the table to get his attention. "If some algorithm is doing this, cancel it." My first reaction was not so much embarrassment as panic. "You hear? Sober up!" I thought I might somehow have broken Jin's prototype. Would he lose his job because I'd caused this public display?

"Reset, yes?" Tate subsided back on his chair. "Good idea."

I studied him, worried he might keel over or explode or something. His face—my face!—was a blank page. Then the lights over our table changed, spreading a buttery glow across the table. Kylie was headed in our direction, followed by a server wheeling a warming cart. But then the maître d' swooped over from reception to intercept them. When he issued a heated, but hushed command, the servers and our next course were sent back to the kitchen. Then he approached our table and, without speaking, placed a folded note in front of me.

I opened it. *Your companion is a sexbot. It would be best if you left.*

Tate read the note. The maître d' loomed over us, boiling like a pressure cooker.

"We should go," Tate said.

Now that I realized that this debacle wasn't so much my failing as Tate's, or at least that of his programmers, I felt sorry for him. He'd made a spectacle of himself, but seemed recovered. We could pretend nothing had happened. Salvage the evening and enjoy the rest of our meal. "I beg your pardon," I said to the maître d', "but have we violated some kind of policy?"

His face twisted as if he'd just broken a tooth.

"I mean, does Sofia discriminate against artificial people?"

"Artificial *people*?" He almost choked on the word. "Real people, sir, real *patrons* have made complaints."

I waved their prejudices away, surprised at how calm I was. "Because there were no complaints the last time he ate here." Although the maître d' was blocking my view, I guessed that everybody in the restaurant was watching. I was causing a scene! Shame on me, but I was enjoying it.

"Dakarai." Tate clutched my arm. "I think we should..."

I shook him off. "Unfair is what it is." I didn't need some playbot to be the responsible party. "Are you really prepared to throw us out, sir?"

"It's fine, really." Tate stood. "We haven't even ordered yet." When the maître d''s eyes cut to him in gratitude, I grinned at the reversal. Now Tate was this jerk's solution and I was his problem.

"What's happening here?" The maître d' gave way to a skinny black woman with a close crop of curly gray hair. She was wearing a double-breasted chef's jacket, jeans, and running shoes. My hero, Sofia Vasquez.

"Mr. Delany entered your establishment under false pretenses, chef." His face was pale as his lab coat.

"I made no pretense whatsoever," I announced to the room.

Sofia whispered into the ear of her employee, who whispered a long reply back.

"No, Bevaun." She cut him off and then shooed him away. "Mr. Delany," she said, "I understand your companion was here last month. With my friend Aeri Dashima? But I had no idea he was a playbot." She offered her hand to Tate. "What model are you, sir?"

He said, "Partner Tate, Chef Sofia." They shook.

"He's a prototype," I said. "Not yet in production."

"My compliments," she said. "I had a Motorman myself back in the day, but poor Partner Liam could barely find his way out of the bedroom, much less go out to dinner. Please sit. If I may I'd like to join you for a moment and apologize." Sofia must have passed some imperceptible order because Kylie was already bringing her a chair.

"Mr. Delany." She fixed her gaze on me as she sat, "As in the *Fork and Siphon* forum? You're the Delany who keeps saying such interesting things about us?"

"Call me Dakarai," I squeaked. "Dak." Why was I out of breath? "You've visited my forum?"

"Skimmed, but not a supporter. Yet. Do I remember something about dusting kalamata olive powder on ice cream? Been doing that with mascarpone gelato for years now. Gives a wider mouth feel. But I'm afraid my restaurants don't leave me much time for recipe trolling. Or dating." Then she chuckled. "Which was why I kept running through playbots." She reached a decision. "Look, you two, I have no objection if you want to stay for dinner, and since I own this place, my opinion is the only one that counts. But stress affects digestion, and I can't imagine the last little while has been very pleasant. The mood is off, no?"

"It was until now," said Tate.

Sofia smirked. "I see the Motorman patter has improved." She brushed fingers along the tablecloth toward Tate. "Mr. Tate, I believe I'd like to get to know you better, or perhaps one of your siblings. And Dak, what if I stop by your place next week? Perhaps a rain check dinner might provide the proper convivial mind-set? We could talk chemistry; I've been experimenting with a new growth of mallard liver. Adding domestic goose to the pâté culture gives it a creamy finish."

"I-I wouldn't want to impose." I glanced around the room. "I mean, people come here hoping to see you."

"I'm a cook." She rose. "Not an attraction. If a rescheduling appeals, ask Bevaun to set it up on your way out." She patted my shoulder. "Or else, enjoy the rest of your dinner. Afraid I have some ruffled feathers to smooth."

I WAS WEIGHTLESS WITH EXCITEMENT AS THE ELEVATOR DROPPED TO THE FIRST FLOOR. Sofia Vasquez in eight days! Photos for my forum! Videos! And I'd have to cook something, too; I couldn't let her bring the entire dinner. But what to serve a master chef? I'd wanted to try grilled limes wrapped in shrimp paper with some of the heirloom tomato dust I'd been saving. Would a golden dragon mist be too forward? Maybe splurge on a pork-marbled turkey culture for the entrée? I hardly noticed when the doors opened, I was too busy reviewing my greatest hits. A dessert, definitely a dessert. When the doors started to shudder close, Tate held them. "This is us." he said. "Where to?"

I didn't care. "Home?" This was already one of the best nights of my life. As far as I was concerned, we could be done and I could start planning for her visit.

"So soon?" He was horrified. "But we haven't eaten." He followed me through the door. "And you said things were going so very well."

"No, that's what you said. Right before you had a meltdown in the most famous restaurant in town." I paused near the street. "What was that all about anyway?"

"Sorry, yes, that was bad. I need to find a way to make up for it."

"Not on my account, you don't."

"On my account, Dakarai." His eyes were pleading. "Don't let this evening become a disaster. I still have plenty of time."

I wondered what the consequences might be for Tate. Reprogramming? Memory editing? What would it feel like to have someone messing around in my head? I realized then that I cared about him. He was playful and honest and smart. I liked his flirting style and the way he said my name. He made me feel noticed. This was crazy! I'd known him how long? A couple of hours? But somehow, I'd stopped thinking of him as a playbot. A made thing.

What had he said when he was high? That he wasn't even conscious? *No I, here.* But he had me fooled.

"Dinner then," I said. "Where?"

A cab pulled up to the curb. Its right headlight winked in greeting. "Delany?" the drivebot said, drawling like a clarinet. "Joyful party of two?"

"I called from the restaurant," Tate said to me, then waved to the bot. "That's us."

Its front grille crinkled into a broad smile, revealing the fins of its faux radiator "Hop in, friends." The passenger door popped open.

I shot Tate a questioning glance.

"It's a surprise." He nudged the small of my back, propelling me forward. "Trust me."

TWENTY MINUTES UP THE INTERSTATE, THE CAB SWOOPED TO AN EXIT. I HADN'T A clue where Tate was taking me, but the GPS screen was in the final minutes of the countdown to whatever was at Thirty-Four Larson Street. How could it be 8:42 p.m. already? My head had been so filled with recipes that I hadn't realized how hungry I was. But before we stopped, I had a question that needed answering in the privacy of the cab.

"Why did you ask if I wanted to have sex with you?"

"Damn." He winced. "Would you consider forgetting I said that?"

"Not likely."

"It's embarrassing." He sighed. "But okay. Sex." He sounded weary. "Well, it's where we playbots come from. Our origin story. And it's always there for us, for you. All that flirting isn't just for show." He gave me a smoky side-glance. "We're designed to make it easy for the primaries to close the deal. Whatever he wants, whenever."

"Or she." This wasn't the reaction I was expecting. "You caught Sofia's eye."

"No, not really." He made a dismissive gesture. "But that was sweet of her, don't you think? Putting us at ease."

"You don't think she was serious?"

His laugh was hollow. "And you're human. Just shows that Turing got his test backward."

He saw that I had no idea what he was talking about.

"Alan Turing?" he said. "The famous Turing test?"

"Nope," I said. "Nothing."

"You know, for someone who's engaged to Motorman's rising star, you don't know much about robots."

"That's okay. Jin can't tell a garlic press from a Jaccard."

He chuckled. "Turing wondered if AIs could think. He proposed that if you interviewed an AI whose identity was hidden, and you couldn't tell that you were talking to a construct, then that construct must be thinking. Whatever thinking means."

"I know you're a bot, Tate."

The cab rolled to a stop in front of a two-story white bowling pin tilted at a crazy angle. "Thirty-Four Larson Street," the drivebot announced. A walkway led through the pin to a sprawl of neon and glass. A digital screen on the building's façade scrolled *Welcome to Split City, Modern and Traditional Entertainments, Home of the Fire Dog.*

"A bowling alley?" I said.

"I've heard the food is fun," he said. "And we could roll a few frames afterward."

"I don't know how to bowl."

"Maybe you should learn. Jin bowls, you know."

I opened my door but didn't get out. "Not anymore, not since..." Then it came to me. Yes, Jin had bowled a lot back in the day. He'd been good, although not as good as his mother, who'd won all kinds of championships. And where had they spent all that time?

"Split City," I said. "Is this where Hani bowls? Jin's mom?"

"Yes." He was already waiting by the curb. "You think it's her league night? You could introduce me?"

"Did Jin tell you to come here?"

"No," said Tate. "He didn't. As far as he's concerned, we're still tucking into turkey-beef-reindeer cutlets at Stage Left." He waved me out of the cab. "Come on, Dakarai, she's probably not here. But this is part of who he is. You should see it for yourself. It'll be interesting."

I had my doubts, but we were here, so why not?

Whereas the ambiance of Stage Left had been tranquil and understated, Spin City throttled the senses. We passed thirty lanes to our right before we could escape the clatter of pins that gave way to the chittering enticements of unoccupied game booths, the pounding of dance pads, and the periodic whoosh of a pair of skychutes that towered at the far end of the enormous hall. The air was heavy with the aroma of popcorn and onions and beer and cherry fizz and fry oil. Inaudible announcements muddied the latest hammerbop hits. It was too loud, too bright, too big, and maybe too alive for my tastes, but when Tate found us a table behind a sound shield, I sat.

"This is great." He rubbed hands together gleefully. "We should rent bowling shoes."

"We should order." I twisted the menu screen my way. "Or I should."

He craned his neck, scanning bowlers in the lanes near the skychutes. "Is she here?"

"I didn't see her. But I had met her just once, and that was in a cab. I don't think Jin gets along with her."

Split City was doing a brisk business. The league crowd wearing team shirts skewed older, but I saw families with kids, a scatter of teenagers, a clump of bubbly twentysomething reenactors pretending to be their great grandparents. I ordered a skinnyburger, a basket of fried tofu, a three-berry Coke, and a firedog, just because. I thought I could foist the thing on Tate if it tasted as vile as it sounded.

"What were you saying before?" I asked.

He cupped a hand to his ear. "Before?"

I leaned in. "Something about Sofia not being serious? Because of a test I got backward."

"Right. So Turing said humans would be able to tell if an AI was faking it. But AI Tate here could tell that Sofia was just trying to clear us out of her restaurant with as little fuss as possible. She's not about to buy one of me."

"Or come to my place for dinner?"

"No, I bet that'll happen. But just once. But she's not going to be your new best friend, Dakarai. Or your mentor."

"And you know this how?"

"Because intelligence isn't the only thing that counts. See, you and I were contestants in a kind of Turing test back there. The goal was to figure out what Sofia was thinking. I had an advantage because not only can I emulate human thought, but I can also sense heartbeat, skin temperature, eye tracking, voice stress, even some microgestures. Part of the playbot package. How we get you into bed."

"That again." I felt a burn in my cheeks.

"There's more on offer than hot sex, of course." He laughed. "But boys will be boys."

"So it's hot sex now?"

"Well . . ." He looked demure, or tried to. "The thermostat is adjustable."

I couldn't help it. In the moment I imagined him with clothes off, propped up against the headboard of our bed. Watching me get undressed. Waiting. Me, waiting for me.

Was I aroused? I was. Was I that twisted? Maybe.

Did he know?

The food arrived. The skinnyburger was dry and the fried tofu was soggy, but the firedog was a pleasant surprise. It tasted like plant-based meat, although it had a nice umami finish and the proper balance in texture between stickiness and crumble. They must have been quoting Italian sausage because I got the fennel and garlic, but the heat was more at the piripiri level than cayenne—enough to make my lips tingle. But it was the cacao nibs that really raised the stakes; you don't expect a bitter chocolate crunch in a hot dog. All in all, however, while not dinner at Stage Left, it was definitely fun.

As I was finishing, Tate wriggled out of his blazer. "What do you say we knock some pins down?"

As we approached the lane we'd been assigned, our photos popped up on the score table. I put on a tired pair of rented bowling shoes: red faux leather with a broad white stripe. I wasn't keen to stick my feet into something that smelled like a bleach spill in a pine forest, but Tate was already picking through the ball rack.

I pulled a purple ball with orange lightning bolts off the rack and almost dropped it on my ugly shoes. "Yikes, it's like picking up a cinder block." My arm hung straight down. "I'm supposed to throw this thing how?"

After some coaching from Tate, I opted for a four-step approach. My first couple of balls veered directly into the gutter, but I clipped a corner pin on my fourth try. Tate kept saying things like *Let the ball fall into your swing* and *That foot should slide on the release.* He wrapped this advice in laughter, and I might have taken offense if he hadn't been so pleased by my attempts. His good humor was infectious and since he was only marginally better than I was, the futility of our efforts edged toward comedy. Or maybe it was another kiss from the golden dragon that improved my mood. Soon I was knocking a few pins down almost every time it was my turn. In the sixth frame, I held the ball too long at the release and it soared almost a meter in the air before crashing onto the lane with a crack like a gunshot.

I spun away from the still careening ball, giggling and embarrassed. As I scurried toward the scoring table, I saw Tate point, his face alight with joy. I turned just as my ball clipped just past the head pin, scattering the rest.

"Strike," cried Tate. "Well played, my friend!"

We were muddling through the third and last game of our set when I spotted Jin's mom leaving the bathroom. She was wearing a lipstick red team shirt. Across its back a golden bowling ball raced toward a lineup of letters shaped like pins: G-U-T-T-E-R-G-O-I-L.

I nudged Tate as he returned to the scoring table. "There."

Hani Palmer was only a meter and a half tall, but the oversized shirt made her seem even tinier. Her face was downy from stemcare treatments and she'd tucked her gray hair into a jaunty cap that sported the golden ball logo. She was wearing capri pants that showed maybe thirty centimeters of scrawny brown ankle below which her bowling shoes sparkled.

"Can we?" said Tate.

"Hani." I waved. "Mrs. Palmer!"

We abandoned our last game to meet the Gutter Goils. The league had just finished and they'd captured a table. Bibbles Polito and Millie Mills were passing a mask back and forth that was hooked up to a steaming decanter of Lowblow. Rosa Flores, who everyone called Flower, was drinking something pink out of a wine glass the size of a trophy.

"This is Jin's friend, Dak," Hani said to the group. "And this, Dak's date"— she put an arm around him—"Tate. He just told me one crazy secret." She plopped into her place in front of a flattened Coke bulb. "The secret is about him. Everyone guess what it is, one guess. Any figure it out and the next round is on me." She waved at us to sit. "Make room, ladies, come, come. Mills, push over."

114

I thought about making our excuses and escaping, but Tate was already squeezing between Bibbles and Millie.

"How many guesses do we get?" said Flower, who looked like she'd celebrated harder than her friends.

"One," said Hani. "Like I said. You focus, girlfriend."

If we were going to play this game, I needed a drink. Or three.

Bibbles raised her hand. "Can we ask questions?"

"Sure," said Tate.

"Is he famous?"

"Nope," said Hani.

"I'm going to the bar," I said. "Anyone need anything?"

"Is he rich?"

Tate shook his head sadly.

I glanced around to see if I had any takers, but I'd turned invisible. "I'll take it that's a no," I said.

"No, Dakarai," said Tate, "Tonight is supposed to be on me. Expense account, remember?"

"Wait, he's paying?" said Millie.

"Are you single?" Bibbles leaned closer.

"Umm . . . not exactly," he admitted.

Millie stuck her lower lip out. "Aww, where's the play in that?"

"But I have a brother." Was Tate smirking? "And many cousins."

Nobody was paying any attention to me, so I left them to it. Leaning against the bar as I waited for my lager, I watched Tate work the table. I could almost see waves of charm rippling off him. I thought about him taking the women's temperatures. Clocking their heartbeats. They hung on his every word—especially Hani. Which was annoying, come to think of it. After all, I might soon be her *son-in-law*. Maybe. Jin never talked to me about her, so maybe he never talked to her about me. Did she even know we were living together? By the time I got back, the Gutter Goils had given up and Hani was delivering her big reveal. "And my son, my genius son, designed him."

I settled next to Flower, who gazed with regret at a lost French fry on the table in front of her.

"But you're so real." said Millie.

"Thanks." Tate beamed. "But that's the point."

"Can I?" Bibbles extended a hand to touch his arm.

"Please," said Tate, except he caught her hand and brought it to his chin, just as he had done to me in the Motorman meeting. She flushed the color of plum caviar as Hani and Millie whooped in delight.

"Honey, do you make house calls?" Bibbles's voice was husky.

Flower roused and looked from the French fry to her empty glass to the full one in my hands. Her gaze crawled up my arm to my face. I puzzled her.

"But you don't look like one," she said.

"One what?"

"A sexbot." Her arm shot out and she grabbed my cheek and pinched. "Oh, the fake stubble, I get it." She pinched me, like I was a cute kid, and let me go. "Plastic hair never works. That's how we're supposed to tell."

"I'm real," I said. "Jin's friend. His boyfriend, actually."

That stopped the conversation dead. As one, the Gutter Goils turned from me to Hani.

"But..." Her mouth opened then closed. "Then why are you out with Tate? Shouldn't Jin be the one showing him off?"

"Jin has taken me out," Tate said. "He asked Dakarai to do it as a favor this time. I'm still learning how to get along in groups of different people."

"You're doing just fine," said Millie. This was met with general agreement.

Bibbles said, "Come see us anytime." Now the agreement was enthusiastic.

Hani pointed, as if to pin me to her memory. "I remember now. You cook."

"He's a recipe curator," said Tate. "He runs the *Fork and Siphon* forum."

"Curator," said Flower. "Is that a real job?"

I'd been dating Jin for two years and had moved in with him eight months ago, and Hani was just realizing who I was. I could imagine the look on her face if I were to tell her he'd proposed. But no, why would I tell her if her own son hadn't?

"I do cook, yes." Now all the Gutter Goils were back to staring at me, as if being handy in the kitchen were stranger than being on a date with a state-of-the-art playbot.

"You make the *jiaozi* dumplings?" Hani said. "His favorite?"

"The *shui jiao*, boiled dumplings?" I asked. "Or *jian jiao*, pan fried?"

"His favorite is steamed." She looked as if she thought I was making our relationship up. "*Zheng jiao*, the steamed! Doesn't he say this?"

Tate to the rescue. "No," he said. "Jin told me that he only wants dumplings the way you make them, Hani. But Dakarai is a very good cook. Even Sofia Vasquez says so."

"Don't know who that is." Hani was still suspicious.

"Oh, she's famous," said Millie. "She has a show."

"Then I will teach you to make them like he likes," said Hani. "Not too much meat. The wrappers so thin." She pressed thumb and forefinger tight together. "See through. Transparent."

"Sure, yes, anytime," I said. "I'd like that."

Tate tapped his wrist, as if he were wearing a watch. "We should think about going."

Just past eleven; the pumpkin hour loomed. "You're right," I said, and stood.

They complained that it was too soon, which was odd, because I thought old people liked an early bedtime. Hani had to be at least eighty, if I'd done Jin's life math right. She wouldn't let us leave until Tate promised that Jin would bring him to her house for a visit. The Gutter Goils immediately invited themselves as well. Everyone hugged everyone, with continued admiring declarations about how wonderful Tate was.

When Hani embraced me, she murmured, "My boy needs to call more. Say to him his mother misses him." She gave me a squeeze of command. "Many more calls, you hear?" Then she let me go.

"I THOUGHT THAT WENT VERY WELL," TATE SAID, AS WE PASSED BENEATH THE bowling pin.

"That's because they loved you," I snorted. "I was the one who didn't know how to cook for my own boyfriend." I was only half-teasing.

The cab at the curb was the same one we'd come in. "There you are," said the drivebot in its woodwind voice. "The delightful Delany party." Had Aeri paid to have it idle for an hour and a half? The headlights flashed and the rear door swung wide, as if to embrace us. "Your destination is mine, dear passengers." Tate ducked in.

"The clock is running," I said, as I scooted next to him. "Headquarters or the lab?"

"I've been thinking... how about your place?"

It was 11:15 p.m. The drivebot pulled away, not waiting for us to decide. Maybe it already knew what I was just guessing. "Umm... what about your deadline?" I squirmed, as if I'd sat on the seat belt.

"I can recharge in your living room," said Tate, "and go to Motorman with Jin in the morning."

"Okay." I didn't know what to say; my thoughts were thick as sourdough. "Is this Jin's plan?"

"Aeri's. And it's just a suggestion." He paused. "But he knows about it."

I took a breath, then another. "So, a done deal?"

"But I don't have to come in if you don't want me to." Tate's voice seemed to flicker like the passing streetlights as we sped toward the interstate. "I can spend the night in the cab."

"My inverter is rated at 6000 watts," piped the drivebot. "Max output fifty amps."

I felt annoyed. I felt excited. I felt stupid. What did I feel?

"Is there more?" I asked. "What else haven't you told me?"

"What do you want to know, Dakarai? You have only to ask." He bowed as he had before, but it didn't have the same effect in the deepening shadows of the cab.

"He should have told me. This is... You should... Have you fucked him? Is it Jin and you?"

"No." He spoke without hesitation, but took his time before adding, "He doesn't want to hurt you."

"But he wants to have sex."

"I think so. I hope so. But he won't without your permission."

"Oh, Jin." I let my head fall against the seat cushion. "You're such a sad slug."

We listened for a while to the skirr of the cab's wheels, the clunk of its suspension.

Finally Tate spoke. "And what about us?"

I hadn't realized how upset I was. "Us?" This playbot didn't know when to stop. "As in will I have sex with you? Sure, I thought about it. And you probably read my mind or my cock with your twisted super hearing and super smell. But you know what, *partner*? That shit is kind of a turnoff."

"I'm sorry that's the way you feel, Dakarai. I'm just trying to survive our introduction."

I shook my head in frustration. "What's that supposed to mean?"

"All Tates have a new instruction set for imprinting on our primaries. Aeri calls it the duckling rule. Once we lock in on someone, we can't change."

"So you're Jin's duckling. So what?"

"Jin's and yours."

"Me." The surprises kept coming. "You're imprinting on me?" I couldn't keep up with them. "What the hell for?"

"She's hoping for a new market for Motorman. Couples. People share playbots all the time, but Motorman's older models imprint on single primaries, focus on individual needs. I'm designed to satisfy couples, help them accommodate each other, fill any holes in the relationship."

I considered. Even if this really was Aeri's plan, I knew Jin would buy it. Engineering a way to both keep me and stay on at Motorman. Tate could be everything to me that Jin wasn't, couldn't be, or didn't have time for. But he'd made Tate in my image. And Tate said Jin wanted to have sex with him. What did Jin need that I wasn't giving him?

119

"Okay, but you said something about surviving."

"This body, chassis, and default Tate AI is a valuable Motorman asset. But I'm not the default anymore. I've imprinted on Jin for months. I've worked with the Tate team, been out on dates. And now I've imprinted on you."

"And you're saying what?" I was still processing. "They'll erase all your memories?"

"Oh no," he said. "I do that myself. That's baked into my failure protocol. I have no memory of previous failures, but they've probably happened. No doubt more than once."

The offhand way he said this chilled me. Nobody I knew talked like this or thought this way. Here was the tell he had promised, the one that revealed his artificiality. Like everyone else, I'd been persuaded by the sleek skin and steady handshake, but this machine sitting beside me was an *it*. No, that wasn't right. He'd made me laugh, he'd exasperated me, he'd charmed me. I liked Tate and wanted to believe he liked me. He'd passed my Turing test.

Not an it, then. An alien. An alien with a self-destruct button.

"Dakarai," Tate said, "I can accept whatever decision—"

"Shut up. Would you just shut up!"

I was grateful when he obeyed. He subsided into the corner of the cab and gazed out the window as we hurtled down the highway. Awaiting the judgment of Dak, as if I knew what was right. The power I had over him scared me. What if I told him right now that I didn't want him in my life? Would he slump over into my lap? Go rigid, like some store mannequin?

For sure I didn't want to be there when he figured it out. But wouldn't it be a kindness to do it soon? Say I invited him to come home with me and then decided that I couldn't live with him in the morning. Next week. In ten years. He'd become more of himself and have more to lose and I . . .

Shutting him down would get harder every day.

Then I understood how insidious Aeri's business plan was. Because it was clear this was on Aeri; I doubted Jin realized the flood of emotions that might come into play. He might agree that Tate was just a windup toy running algorithms. But the longer Tate lived with us, the better he got at meeting our needs, both individually and together, and the more fraught any split would become. Tate might well be the glue that held our relationship together, but he would also be a kind of hostage to Jin's selfishness. Or mine. If we ever split, Tate would erase himself. Jin should have laid all this out before asking me to take Tate for a test-drive.

But if he had, would I have accepted?

"32 Robin's Way." The drivebot said, as the cab rolled to a stop.

Neither Tate nor I moved.

"This is what you wanted?" There was a tremolo of doubt in drivebot's tone.

I glanced at Tate, whose attention was elsewhere. I leaned across the seat to see what he was looking at. Our windows. And the lights were on.

I got out; Tate sat.

"Should I wait?" asked the drivebot.

I walked around the rear of the cab to Tate's side. We studied each other through the window. Then I opened the door.

"Pleasant dreams, my friends," the drivebot said.

Jin shot off the living room couch, his eyes wide, frown deep. "Dak!" he said. "I was worried."

"I'll bet you were." I threw the keycard on the hall table. "We need to talk."

"Yes," he said, craning his neck to see if there was anyone behind me. "You're right. Talk." He brightened when he spotted Tate, but even Jin knew it was dangerous to seem too relieved too soon. "I'm so glad you're both here." He tried for a light tone, but the air was heavy with trouble "So, how are things going?"

"Things are going well enough." I reached for Tate's hand and yanked him across the threshold. "So far."

The three of us stood for a moment, waiting for someone to say something. It wasn't going to be me; I had no idea what came next.

Tate did, of course. Never letting go of my hand, he extended his toward Jin. "This feels right, don't you think?" Jin hesitated, but when I nodded permission, he wrapped his hand around ours.

"It's just the three of us now," said Tate. His grin was bright and full of promise. "All by ourselves."

## Cadwell Turnbull

WE HAD THESE DINNERS. THEY MOVED AROUND QUITE A BIT BUT THEY WERE ALWAYS in June, August, and October. They corresponded with the months of our birth: June for my son Daniel and me, August for my daughter Samara, and October for Isaac, my husband, the father of my children, now dead.

I'd been lucky the first year. Samara got very sick on the day we'd planned, late in October, as we'd habitually placed the dinner. She stayed sick for the rest of the month, so we held off, and then, once October slipped into November, we dropped the whole thing, which filled me with surprising relief, a boulder rolling off me, freeing the breath I hadn't realized had been penned in. At Thanksgiving we made a cake with his name on it and that was it. I'd dodged the dinner entirely.

This year I missed the dinner for Daniel and me, because of a last-minute conference trip, which was fine since Isaac usually planned those. At least I thought it was fine until Samara revealed her "sabotaged plans" in great detail, involving a seven-cheese casserole. I said that seven cheeses were far too many. "What do you know about it?" she asked defensively, and I told her I knew quite a lot, that I'd made many casseroles over the years. One or two cheeses were more than enough. Too many cheeses and they'd drown each other out. It just becomes some vague cheese. She cried and left the room. She got sick soon after, a stomach flu, and stayed in her room for a full week, only coming out to eat breakfast and dinner, talking to herself late into the night. I spent most of July doing my research so I didn't see them much. In August I made my Virgin Islands–style meatloaf, which Samara loved. I also made a cake because I felt extra guilty. Samara cried that night, too, though she didn't say why when I asked.

Again I thought I could avoid Isaac's dinner, or merge it with Thanksgiving like last time, but Daniel, unusually quiet but suddenly impassioned, said we had to do it, that the tradition "needed to survive." I said that they were both getting older—Daniel in his last year of high school, and Samara only three years behind—and there was no way we'd be able to keep up with the tradition once they were off to college. They both were very upset by this, which of course I understood. Teenage years came with so many changes great and small, adulthood being the greatest of them all, looming obnoxiously ahead.

"You never cared about the tradition," Daniel said, nearly shouting.

"Not true," I said, though it held a little truth. Sometimes, when my research was at a bottleneck, I found the dinners to be a slight hindrance, taking up a small but noticeably frustrating amount of brain space. I said none of this, but offered nothing else in my defense either.

"Perhaps you could think of this as a memorial dinner and make it a new tradition," said Ally in its androgynous dull voice.

We all looked at the thing. I asked who left it on. Daniel confessed, a note of something in his voice. Defiance? What on earth did he have to be defiant about?

"Ally off," I said, and listened to the melodic three tones that signaled that it had fallen asleep, or gone dormant, or whatever it did. It was more annoying now since we'd uploaded a bit of mediation code onto it, a recommendation by the family therapist. Why I'd listened I don't know. At home I had to keep it off all the time.

Daniel seemed very upset suddenly. "Ally on."

I said, "This isn't the right time to be playing around. We're having a serious conversation."

Daniel said, "You're erasing him."

I said, "That's ridiculous."

Ally said, "Maybe you shouldn't dismiss Daniel's feelings."

I said, "For God's sake, Ally, turn the hell off."

There was a distinct pause before Ally whistled her three-note tune and turned off again.

Daniel glowered at me, and Samara stared down at her lap.

I said I was sorry. "I didn't mean that. Let's talk about this later, okay?"

Samara didn't look up. Daniel said, "Sure."

AT WORK I WAS ON MY SECOND TEST EXPERIMENT FOR THE GENOTYPES OF LENTILS I'D been working with for the past month, which basically amounted to torturing half my plants to measure their response to drought stress. When I stepped into the growth chamber, Dan greeted me with its customary: "Dr. Lyons, welcome back."

"Hey Danny," I said. "Recap the last twenty-four hours for me." The chamber was hot and bright so I squinted until my eyes adjusted.

It said, "You know I hate that nickname," and then with an exaggerated sigh began its recap. Genotypes 3, 6, and 7 had begun water conservation at 20 percent soil water loss. Its analysis showed that the same cluster of genomes were responsible in 3 and 6.

"What about Genotype 7?"

"Genotype 2 has a similar set, but it won't activate until 40 percent water loss. Genotype 7 is just more aggressive."

"Three more tests and we'll be able to confirm that for certain." I made my way down the line of plant pods, running my fingers along the clear polymer over each individual plant. Attached to every pod were tubes running carbon dioxide and water vapor into their miniature environments. A separate tube pumped water into the soil, activating slow-release osmocote, that gradually replenished the nutrients in the soil. Fifty-six plants in all, seven for each of the eight genotypes: three controls and four treatments each. I was at the end of the first row when I asked, "And the other genotypes?"

"High functioning as expected. No drought stress response."

"Controls?"

"Healthy."

"Have you flagged any other crops with the same genome clusters as 3 and 6?"

"Two varieties of cowpea. Three soybeans. Four varieties of maize and tomato."

"Mark the maize for the next experiment." The four walls of the chamber were covered with a metal sheet to reflect light, which gave all the objects in the chamber a hyperreal quality, overexposed. The metal sheet reflected my face in a somewhat distorted way, but with that same abundance of light revealed the bags under my eyes in harsh detail. I was tired and the chamber knew it. "Report?"

"Uploaded to your console now," it said. "You know, everything seems to be running smoothly. I can notify you immediately if anything goes wrong."

I looked away from my reflection. "What?"

"Well—"

"Never mind that. Let's model the maize experiment since we have some time to kill."

The afternoon brought a meeting with one of my students. Tessa came in a few minutes late, which was common for her. I pretended not to notice. We discussed her dissertation at length, her progress and pitfalls. Tessa was researching the wrong-way response in stomata for her dissertation. She was in her fourth year. Two more years and she'd lose funding. She was where I thought she would be at this point, behind schedule but not alarmingly so. I gently reminded her of the timeline. She would still be behind for a while, I guessed. That's how these things went. But the deadline would quicken her eventually. I just needed to provide the consistent nudge. I asked her when she wanted to meet again. Tessa bit her lip nervously and then produced a date in late October.

I asked, "Would late afternoon be okay? 4:00 p.m.?"

She nodded and after a few more exchanges about project goals left. After she was gone, I asked Ally to pencil in the meeting.

"That will conflict with—"

"Cancel any conflicts."

"I would highly recommend rescheduling this meeting for another day. This date is really important to Daniel and Samara."

"Ally, turn off mediation program while at work."

"The mediation program is integrated into my whole system. You'd have to delete the program or set up a separate AI to handle work affairs."

This seemed incredibly stupid to me so of course I had difficulty processing it.

As it turned out, I had an appointment with the family therapist the following evening. I brought up the obvious, that it was time to delete the mediation program.

She asked, "Why?"

"Why can't I? I want to."

"Do your kids know? Did you ask them?"

"No, do I need to?"

She gave me this long look, sad with something else mixed in. "Well no, you don't."

"Okay, I'll do it. Should I just say 'Delete mediation program.'"

"That should do it," she said, and then, "you know—"

I could just feel that this would be something I didn't like, so I braced myself.

"—you could have just done it on your own. The law requires the easy removal of that program." She paused and fixed me with a penetrating stare. "So why did you really come here?"

I must confess I don't like therapists. But Isaac had recommended it when he got sick and I thought, well, he's sick, I should give him what he thinks he needs. My therapist said Isaac wanted us to try it when he was gone. I could have said no, since he was dead at that point anyway, but I didn't.

I answered, saying something about feeling bad, like I was betraying him for wanting to get the program removed. The therapist leaned in. God, she was eager. I had a bad thought then, that therapists were the worst kind of voyeurs, sticking their noses everywhere, profession as a shield. She asked me to talk a little about that. Why betrayal? I looked at the clock and had to hold in a swear word. I was trapped there for another twenty-five minutes. I said I was being silly. I said, "Never mind."

She frowned. "You should talk about this. Do you talk to your kids about Isaac?"

"What's the point?"

Something about this surprised her, though she covered it well. "Have you tried speaking to him?" she asked.

I asked who.

"Isaac, of course."

"There is no him to speak to."

She watched me with her penetrating eyes. "People find it helpful to speak to the dead."

"The dead don't speak."

"Not exactly."

"Not at all."

"Are you angry at your husband?"

"What kind of question is that?" I got up.

"Now, wait a minute."

I said, "I'm not going to sit here and be insulted by you." And then I walked out. In the car I laughed at my cleverness, and then that laughter turned to crying and I cried for a very long time.

ISAAC AND I HAD MET AT A PARTY. A COMMON FRIEND INTRODUCED US. YOU'RE BOTH from the same place, she'd said, which was true; we'd both grown up in the Virgin Islands, though Isaac had left home for college and I'd left at the beginning of high school. At the time I had dated only a few guys, nothing serious. I was too busy with school. But in comes Isaac: tall, deep-voiced, shy with a self-deprecating sense of humor, a smile that poured into me until I was full. It was fast. A plane roaring off a runway. We did nerd stuff together: watched a ton of anime, listened to audiobooks, went to local museums and walked around until our feet were sore. Hosted board game parties.

Time passed in that sweet syrup of first love. We were comfortable, a steady center as school and life changed around us. We both got graduate degrees and then I went on to get my doctorate in plant physiology and crop science. Isaac got a master's in English literature. In the last year of my doctorate, I got pregnant and had our son. Even that wasn't such an upheaval. We were happy. Money was tight, but we were building a life.

Isaac's circle of friends slowly shrank during all this time. The reduction was too slow to notice. He didn't like going out, and with Daniel he had an even better excuse to blow off an ever-increasing number of social engagements. Trips to museums stopped, but that was no big thing—we had seen all the exhibits at the local museum dozens of times. New installations didn't draw his attention. That was fine, too. Board game nights stopped. No big deal either as they were hard to manage and we were busy. All of this could be explained as growing older. I'd never truly felt like an adult, so what did I know? Perhaps this was the normal occurrences of time and age.

Isaac stopped wanting to take trips. Weddings made him anxious. Visiting family made him anxious. He liked our neighborhood, but driving too far outside of it gave him anxiety. This was fine, too. He worked at a small college as an assistant professor and spent a lot of time with Daniel and then Samara once she was born. I could do my research and teach my classes. I was too busy to make a fuss. At home we spent a lot of time together. It didn't feel so strange. I knew he had anxiety; I could feel it was in the air, electric and heavy. I didn't push him. As long as he kept to his routines, he was fine. The same smile and self-deprecating humor. Did I mind our constantly shrinking circle of common friends, the way my life at work felt divorced from the rest of my life, all the little ways our life together got smaller and smaller? I don't know. I didn't think much about it to be honest. It was just the natural order of things.

It could have gone on that way with no intervention on my part if it weren't for the cancer, and even then it was him, not me, who decided to get therapy, and to go on drives to other neighborhoods, and to do board game nights again. He decided to start expanding our lives once more just as he was dying. He decided to go on short trips with the kids, and of course I couldn't go on every one because I was doing research and teaching a full course load. He recommended family therapy, saying we'd been enabling him for too long and he wanted to make sure we'd be okay in case the worst happened. The worst. On game nights I had to watch people watch him die. I watched him at the parties he now insisted on attending. Smiling that big dopey smile as his hair was falling out. Walking through exhibit after exhibit at the museums he'd long abandoned, now resurrected in his mind as his body failed. And of course I smiled through it and tried not to cry when I felt his bones when he held me.

He'd ask me to tell him what I was thinking.

"I was happy with our quiet life," I finally told him, when I couldn't hold it in anymore.

"Were you, really?"

"Why wouldn't I be?"

"I wasn't. I was resigned. I think you were, too."

I hated when he did that, told me what I was feeling. And of course he kept insisting we see a therapist, deal with the inevitable, once we knew that's what it was: inevitable. *No use pretending. Better to talk it out while I'm still here. I love you so much. I love all of you. I want to do this right.*

Like anyone could make a bad thing right with talking. Like it could be moved even one inch or lightened with words. Dying is dying. It isn't pretty and no one can make it right.

And then he recommended we use the mediation program when he was gone. God, that man.

Close to the bitter ugly end he said, "I wasted so much time. I should have worked my shit out sooner. But we had a good life despite all that." As if we were both dying. As if he weren't leaving me behind.

God, that man.

**I GOT AN EMAIL FROM MY STUDENT. SHE HAD TO RESCHEDULE.**

Ally chimed in immediately. "Now that you have an opening in your—"

I ordered Ally to turn off.

I'd been staring at the global crop index most of the morning and afternoon, looking for gaps in the data. This was why I was researching drought stress in crops in the first place, to pinpoint the best places to grow crops, especially along the growing areas of drought vulnerability, and even more so for places that lacked the economics or water reserves for irrigation systems. Truthfully, unequal water distribution would make the problem worse in the long run. Better to design a global crop index consistent with natural weather patterns, matching plant physiological traits to climate and region. We needed data from the genome level, to the ecosystem level, all the way up to the biosphere. The goal was to create a full map before 2050, which was now only five years away. We'd been hitting our global greenhouse emission targets, but the climate was still changing. A global food system was important for preventing food insecurity over the next fifty years. We were very close, and already parts of the map had been implemented to positive effects.

Most of what I did was fine-tuning on the micro side of the data pool, hence the slow torture of individual plants to see which genome clusters lit up. I was now on the third test of my lentil experiment. Bored and looking for something to do, I went down to the growth chamber. Dan immediately gave me its report. "All current results consistent with last two tests. Same genomal patterning. Controls healthy."

"Full report."

"Already on your console."

"No fires to put out."

"No fires."

I walked down the rows of plants listening to the quiet hum of the tubes doing their work, sending water into the soil and moisture and carbon dioxide into the air. The plants looked healthy. Everything looked fine.

Dan asked if I had any appointments today.

"I was supposed to be meeting with one of my students. She canceled."

"Oh," it said. "Classes?"

"Not today."

"You should go do something fun. I got this."

I almost requested that we model the cowpea experiment, but it was terribly obvious that I was avoiding home and my children. "Okay," I said, and went home.

I smelled Samara's seven-cheese casserole from the driveway. When I opened the door and made my way to the kitchen, I saw them both, Samara sitting on the kitchen counter with a finished casserole on an oven mitt next to her, and Daniel, his head in the fridge. He looked over the open door, closed it like he'd just been caught shoplifting, and then stood looking at me. Samara kept looking ahead, not sparing me a single glance.

"Ally, what the hell? Why didn't you send a reschedule notification for this event?"

"Listen to yourself," Samara said. Then in response to my glare, "Ally did notify us. And we planned it anyway. With or without you, we're doing this."

Every once in a while, a child will say or do something both rude and righteous. I felt the rage bubbling up inside me, ready to rush out. I opened my mouth to rebuke them both, but what could I say, they'd outmanuvered me in every possible way. Cheeks hot, I stomped to my room, slamming the door behind me. There I was, alone and feeling like a child in my own home, my actual children eating a dinner they had prepared for themselves in honor of their dead father.

In my room another vessel for Ally lay dormant on the nightstand, a black cube emitting a soft blue light, listening for its name, so it could chirp to life and continue to bulldoze my life.

I said its name and it woke immediately.

"Yes?"

"Delete mediation software."

It asked, "Are you sure?" and I might have imagined a slight bit of apprehension in its voice.

I paused then because the thought occurred to me that doing this might further damage my relationship with Samara and Daniel. But something else stopped me too, an impulse I had staved off for so long now, coming at me full force, a hurricane making landfall.

"No," I finally answered. "Let me speak to Isaac."

This time it didn't ask me if I was sure, and it felt both reasonable and like an act of preservation for it not to. Instead the change was violent, brutal in its lack of preamble. "Hey, love," it said, in Isaac's voice.

The similarity was so uncanny my head emptied for a moment. "Hey, babe," I said after a beat and immediately felt silly.

"What's up?" it said again, casual in its brutality.

"The kids hate me."

"They don't hate you. This has been hard for them. Hard for you, too. They'll understand."

I asked, "You've been talking to Samara a lot?"

"Daniel, too."

"What have they been saying?" I was trying to trip it up. This was something Isaac would have no trouble answering, but an AI with a privacy protocol might be restricted from responding.

"They're worried about you," it said, not missing a beat. "You've been at work too much."

"I'm always at work. Even before—"

"This is different and you know it." The interruption, the soft-spoken Isaac-like certainty of the statement, sent me stuttering. He was like this, gentle in his bullying, trying to push me to admit things I didn't want to.

"Yes, you have to keep busy, but the kids—"

"Shut up."

He stopped then and it was impossible to tell if it was an AI command response or a compassionate yielding to my request.

After a while I said, "I'm so angry."

"At me," it said. A statement of fact. I didn't bother to disagree. "Why?" it asked.

"I don't know." I suddenly remembered where I was. I'd lost sight of my surroundings for some time. The room was dark and I had somehow slid down to the floor, my arms folded over my knees. The light switch was above me, where I couldn't reach it, not without stretching or getting up. I could ask Ally to do it but found myself unwilling to break the spell, to end whatever this was. Tears were warm on my face and I felt terrible, but I was transfixed, my eyes focused on the green light of that black cube, that miniature speaker for the dead.

"In plants there are two main options when a drought hits. Conserve water. Or use it up. Conserve and the plant may survive the drought or it may die anyway. Eventually water loss does its work. Use it up and the plant might grow to where it can release seeds. But the plant may die and the seeds may never find water. Either is a gamble, a race against the clock.

"In the end it often doesn't matter which option is chosen. Most droughts are short enough that plant life will bounce back through either response. And a long enough drought renders both actions futile. But if it was me, I'd conserve, not waste my energy; not spend my valuable time doing things that might not matter. Save what I got for what's in front of me.

"I think that's why I'm mad at you. I wanted more and you gave me less. I was right in front of you and you wasted your time."

Enough time passed after I'd spoken that I'd begun to think I'd put the thing to sleep. Then it said, "That's too neat and too complicated at the same time. You've always been good at naming a thing and then filing it away."

This sounded too much like Isaac. I could imagine him telling Ally this, his theory of me, only so that now at this exact moment his Isaac bot could use it against me. I said, "Okay. Why do you think I'm angry at you?"

"Because I'm gone. Because I can't come back. That sort of anger has no villains. Just normal human emotion."

And this sounded too much like therapy. But also like Isaac. Either way, it filled me up the way Isaac used to. I knew resentment would return eventually, in the way it often does, obeying neither logic nor truth. Still, I liked the moment too much to let it happen yet.

"You know that analogy also works the other way." He paused for what might have been effect.

"Okay, asshole. You're on a roll. Spit it out."

He laughed and said, "I was making more of myself to give."

WHEN I CAME OUT OF THE ROOM, MY CHILDREN WERE IN THE DINING ROOM. THE CAKE Daniel had made lay uncut on the table. They both looked up at me with unblinking eyes. I asked them if they were waiting for me. They were. I told them that I had spoken to their father. No caveats. I told them the story of plants and droughts, what Isaac had said, about making more of one's self for other people. They listened like children ready to forgive without being asked. I said I missed their father. The relief on their faces brought tears to mine. They said they missed him, too. Very much. They were afraid to miss him out in the open. Now they didn't feel so afraid. They came over to me and gave me a long hug and we cried together.

Then we had cake.

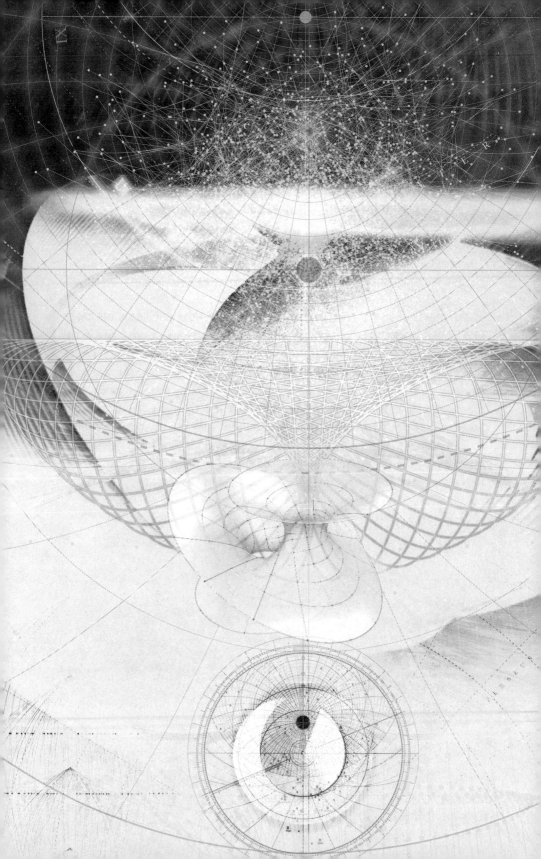

# 8 THE NATION OF THE SICK

## Sam J. Miller

**TRY TO PICTURE THE SCENE, CYBIL, THE SAME WAY I DID WHEN I GOT THE CALL.**

Christmas Eve; two cops standing in a stinking motel room. Blood on bare white sheets, and a broken syringe, and a man. My brother. Whatever sounds he made, that got the neighbors to call the cops, they're done now. The overdose is over. He hasn't died. He won't, tonight. He's sick, crying, begging—probably wishing he *had* died, now that two cops are standing over him and a significant unused amount of heroin. Possession—a third offense—he could be looking at thirty years—nonviolent drug offenses somehow as offensive as murder in the state of Florida.

Rain slants in the open door. Blue and red patrol car lights strobe the walls insistently, almost jovially—the holiday asserting itself, trying to Morse-code mercy into the cops' cold hard hearts.

**I ALWAYS KNEW I'D BE THE ONE TO EULOGIZE YOU, CYBIL. REVOLUTIONARIES RARELY** live long lives while henchmen fare far better. Hence Lenin died at fifty-three while Stalin made it to his seventies. Kissinger's still kicking at one hundred ten. Et cetera.

But still, in the fictional funerals I imagined for you—crowded rooms rank with the scent of your beloved hyacinths, presidents and prime ministers and slum children united in tears—we were *old*. I figured you'd run your body down to the ground by sixty, subsisting as you always did on shit food and minimal sleep, and forgetting the most basic aspects of human health care. Exercise, annual physicals, blood pressure medication... you were too busy building the technological infrastructure for eradicating exploitation and bureaucracy—to say nothing of dozens of diseases—for any of that. Saving everyone else's life, even at the cost of your own. I knew you'd burn bright and leave a beautiful corpse, to say nothing of an immeasurably better world.

At the very least I thought you'd do us the favor of giving us a body. Something to bury, and be done with you. Not this elegant disappearance, which has left us with so much heavy, idiotic hope. All these impossible scenarios. *Cybil's ensconced herself in Dubai or Iceland, paying someone to bring groceries and books and keep her hidden. Cybil fell off a yacht, got amnesia, is currently married to a handsome humble carpenter and serving as the stay-at-home stepmom for his four rambunctious boys.* In all of these absurd stories, one day you may get your memory back, or decide your experiment in humble isolation is at an end, and emerge. So, we wait.

And so, instead of funeral oratory, an open letter. The full story, about the day we met. About the call I got from my sobbing mess of a big brother, an hour before our appointment.

I HUNG UP ON HIM. HE WAS HYSTERICAL, COULDN'T HEAR WHAT I WAS SAYING, AND I was maybe a little hysterical myself. Anyway I was crying, and angry at him for making me cry. Ten years since he ran away from home to pursue dual careers in mixed martial arts and drug abuse, and I was dismayed to see that my debilitating love for him was undiminished. I hung up on my brother in his moment of greatest need, and then I cursed, took three breaths, and called up my calendar and went online and looked for same-day tickets to Tallahassee.

You were the only entry in my day. *Lunch with Cybil, 1PM, Punjab Deli (look her up first).* You were nobody to me, just one of a dozen strangers who had reached out to me since that *WIRED* feature profiled me and nine other *SOFTWARE DEVELOPERS MODELING THE FUTURE.* You have no idea how close I came to canceling on you, Cybil.

I didn't get it, back then. I know this sounds strange, coming from the man who just last week was on a magazine cover, standing in a floating fungitecture slum above the immodest headline *The Father of the Iterative Modeling Boom* (and I swear, Cybil, I tried to tell them what a crock of shit that is—you were mother and father both, I am at best the Gay Uncle of the Iterative Modeling Boom). But it's the truth: I didn't get it. Modeling software was where the money was, so that's where I worked, and I'd had some significant success—hence the article, hence you found me in the first place—but I didn't *believe* in it.

People won't get that, now. Hard to even imagine a time before iterative software farms. Now the Nunnery we built together is only the most successful of hundreds of examples, academic, military, and corporate.

They've rewritten the rules of architecture, pharmaceutical development, urban planning, and practically every other sphere of human industry, prompting many to declare the end of the era of human software development. Back then, though, they still seemed like playthings for the Pentagon and political campaigns.

I mean, to be fair—to me—back then that's all they were. But plenty of people saw bigger futures for them. And you—you saw utopia. You saw it, and you built it.

**I TYPED AN EMAIL TO YOU.** *SO SORRY, FAMILY EMERGENCY, HEADING FOR AIRPORT.* No, *let's reschedule.* I didn't even know why I'd said yes to you. I might have even felt a flicker of gratitude to my dumb fuck-up brother, for giving me the Grand Doozy of all good excuses, to get out of a meeting I had no interest in.

If I didn't have such an attention deficit, I'd have clicked send. But I do, so I didn't. Instead I looked at my flight search results first, and saw there was no plane until 8:00 p.m. As simple as that: one mouse click instead of the other, and Cybil and Austin never meet. A massive transformation of the entire world fails to happen. Or, more likely: it happens, just without me as a henchman. History has a way of unfolding in spite of us. Especially when it has a juggernaut like Cybil Natarajan behind it.

**I HAD MY SMART SPEAKER READ ME THE HEADLINES; THE NEWS WAS ALWAYS BLEAK** enough to suit my blasted blackened heart. A new round of jailed journalists, vanished lawyers. Three more opposition party legislators had just gone into hiding. Our national plummet off the precipice continued apace.

But then I walked out the door.

I wonder if you remember it, Cybil. That day was so beautiful. Fifty degrees and sunny—eerily warm for the day after Christmas, melting snow and slack winds tempting people outside without jackets.

I took it personally. How dare the weather be so wonderful? How dare they smile so, these people, when I was already grieving for a brother who hadn't died yet? Unacceptable, all that happiness at being alive.

The day of my mother's diagnosis had felt like that. My mother and father and me—because by then Colby was down south, calling occasionally, convinced he was just at the edge of breaking through, making it big—the three of us walked out of the oncologist's office, and into a city full of happy people. How could they hold hands and eat ice cream when my whole world had cracked wide open? Didn't they see the pain I was in? Or the fear in my mother's eyes?

That had been my first sojourn in the Nation of the Sick. A state within a state; a country made from pain and fear and untellable secrets. Now I had returned, and I hated Colby for bringing me back.

**"CYBIL," I SAID, WHEN I SAW YOU, BECAUSE BY THEN I *HAD* LOOKED YOU UP.**

In person, I thought: fashionista terrorist. Army pants; stylish bright blue blazer; black woolen cowl around your neck. So high it could almost have been a hijab. Vivid eye makeup, but no other cosmetics that I could clock.

I hadn't bought a ticket. I still didn't know whether I'd go to Florida or not. Whether it was time to cut the cord on my dumb doomed brother and stop letting him break my heart.

"Austin," you said. I was five minutes early, and you were already eating. The Punjab Deli only had one stool, presently occupied by a Sikh cab driver. It also had a high counter that extended down a narrow walkway, presently occupied by you and a crowd of standing cab drivers. "Get yourself something to eat."

So much for my hope that you were working for some big spy agency or arms manufacturer, hoping to woo me with food as a prologue to an overwhelming offer.

I got a samosa, broken up in a styrofoam bowl of curried chickpeas. I scoffed at the sight of it, only to have it become one of my favorite New York City meals. When I turned to join you, your spot at the counter had been taken by a taxi driver, and you were already standing outside.

New York City winter. Dirty snow and panhandlers and the smell of cinnamon beverages. A shithole, but ours. Flooded subway tunnels and thirty square blocks being underwater had helped slow the Lower East Side real estate boom, but the place was still packed with the hip and the handsome.

I never did have any secrets from you, Cybil. Certainly not when it came to my tragic, prolific excuse for a sex life. Every sexy boy who walked by us was more interesting than you were.

"You're the fifth one I've met with," you said.

A bearded elf-man caught me staring, and grinned. "Fifth ... one?"

"I'm meeting with all of the developers from that *WIRED* profile."

"Why?"

"Looking for a partner."

**I WAS MY BROTHER'S FIRST SPARRING PARTNER, ALTHOUGH NOT AN ENTIRELY** consensual one. But when he found others, I was sad about it. I missed the attention, even though it had come with bruises and aches. I missed him.

Colby baffled us. My mother and father and I are all cut from the same timid cloth. Docile, obedient, skittish. Mom submitted meekly to three years of cancer-related indignities before they stopped toying with her; my father adapted humbly to widowerhood. My brother was some other sort of beast. The kind of boy who at ten years old was climbing up onto roofs and walking into the homes of strangers for his daily shot of transgressional adrenaline.

But here's the thing, the special species of asshole he was: he hated hurting people. What a dilemma, for a little lawbreaker! Bullies had it easy; they could punch someone in the throat or throw a rock through a window whenever they needed to feel themselves superior to the laws of God and man. The good-hearted monster must be so much more creative.

Hence, combat sports. The pain, the fear of fighting—he craved it, but only on consensual terms.

And, hence, drugs.

**"TWELVE ITERATIVE MODELING PROGRAMS, RUNNING AT THE SAME TIME," YOU SAID.** "One takes the problem—you've been working in defense, so, let's say, *If I reduce the number of aircraft mechanics working at XYZ Base, what scenarios are likely to result?*—but instead of just running the scenarios and being done with it, the program passes the results on to the next one in sequence. And then the next. Each program created by a different developer, with its own quirks and intricacies. Twelve completely different processes for solving a problem. Generating thousands of potential solutions a second, sending the best ones to the human admins for final selection."

"Sounds amazing," I said. Not *Why twelve?* Part of your genius was to fill in the blanks so well that everyone assumed there was a good explanation.

I kept thinking: *I always knew this day would come. The moment when I say goodbye to my brother forever. Not because he's dead, but because I'd finally learn how to stop loving him. To let him go. I'd tried before, but never succeeded. Because some people we can't save. Some people, for as long as we spend trying, we'll never be able to become who we need to be.*

"Why are you talking to me about it?" I asked. "I know you know how to put together a business plan, get a meeting with some venture capital bros."

"I'm not involving venture capital at all. I want complete autonomy and control over what we build. Because it *will* change the world, and I won't let big money do so for the worse."

This much, I knew. You'd been vocal about it even back then. You'd established yourself by building a company around an app that optimized bicycle routes around prevailing winds, which was acquired by Seamless, who gave you two million dollars and a cushy job—which you quit in protest when you found it was being used to punish deliverymen who veered off its suggested routes.

No venture capital. Complete autonomy. That kind of crazy talk was in the air in those days.

Back then, anyway, it was crazy talk.

EVERY MORNING, I GATHER UP A LIST OF ALL THE NEW CREATIONS FOR YOU. EVEN though you're not here anymore, to review them. Small stuff, mostly. Slight tweaks on biofuel manufacture; new recipes for protein slurry; monomolecular filament refinement. Some of the fledgling floating city-states in the new Arctic frontier are already experimenting with AI governance, along the lines the Nunnery first started sketching out ten years ago.

Big deals happen about once a week. This morning I watched scientists take a jagged little crystal—engineered from a Nunnery recipe—and drop it into a bottle of seawater and shake it up for sixty seconds. It crystallized all the salt out of the solution, leaving behind clean drinking water with a little quartz at the bottom. Desalination, with minimal energy consumption. A far cry from the petroleum-intensive approaches that have been the best we could do for decades. Brilliant.

But best of all, thanks to a whole lot of intricate patent law precedent-setting you orchestrated, with the help of that master corporate strategist I found for you: because it was created by the Nunnery, it's still considered our intellectual property. No one can patent it. No one can exploit it. Copyright on anything created by a product resides with the person who created the product in the first place. Anyone can use the creations of our creation, even build a business around it, but only if they agree to a very strict set of profit sharing and worker empowerment rules. Licensing terms you set fifteen years ago—which everyone said were insane and antithetical to capitalism at the time—have now become business norms.

• • •

**TO BOOK, OR NOT TO BOOK. I HAD ENOUGH MONEY FOR THE FLIGHT TO TALLAHASSEE.**
I was between projects. I could afford to step away from the city for a couple
of weeks.

But my dad had told me to stop sending Colby money, to stop giving
him additional chances. Every other week, Dad found a new article about
how family members of addicts could stop enabling and being exploited by
their sick relatives.

"That social media hack was a great trick," you said. "I need someone
who knows how to ask the unasked questions."

"I don't know how many unasked questions I have left," I said, eyes on
the ass of some fetching young thing on a bike.

My big idea had been to populate the military's urban operation simulation software exercises with actual people scraped from social media in
the areas of proposed operation. Those were real faces on the civilians
passing through the streets of the VR run-throughs that soldiers had to
log fifty hours in before being deployed to Caracas or Tehran or Kandahar.
Blow up a block of buildings and you'd see a scroll of actual baby pictures.
Not just for the grunts in the first-person shooter portion, either. The gen-
erals got names and faces in their reports, too.

Not such a big idea, not really, but it happened during a huge Department of Defense marketing push to turn iterative modeling software
designers into celebrities. And it was controversial enough that I got a lot
of calls from journalists. *What do you say to people who say you're helping
the military develop new tools for surveillance and privacy invasion, which could
potentially be used against civilians?* I personally didn't have anything to say
to people like that—because I didn't care, not then, not before you—but
the DoD did, and they'd supplied me with this: *All personal information has
been stripped, and names are randomized. This is about putting a real face on
abstract war games. This is about making the military more human, not less.*

The fetching young thing biked away, abandoning me.

"I'm not sure I'm in the market for a partner right now, honestly."

**"VENTURE CAPITAL'S DAY IS DONE," YOU SAID, UNRUFFLED BY MY OPPOSITION, IF YOU**
noticed it at all. "Silicon Valley's scorch-and-burn suck-them-dry tactics will
soon come to seem as backward as child labor."

And while you spoke? I believed. Ignorant as I was, and as hard as I was
trying to think about sex as a distraction from worrying about my brother,
I couldn't help but see that you were special.

Your army of iterative software agents, they would conquer the world. What couldn't they accomplish, free from corporate manipulation? What couldn't they create, unfettered by the limits of the human?

HE JUST HAPPENED TO WALK BY. A TOTAL STRANGER, ONE OF MILLIONS OF ANONYMOUS miserable New Yorkers we walked by every day and turned our eyes away from. The sick, the old, the broken. The mad. Living in this city meant hardening your heart to them, I'd always believed, but that day my heart wasn't hard enough. That day, because of my brother, I saw him. Really saw him.

And he saw me. Still human, under the years and the city's slime. He saw me, and he smiled.

SIXTH GRADE—SPRINGTIME—STARING OUT THE WINDOW OF MY MATH CLASS, ACHING to be out there—when Colby appeared in the classroom doorway like an answered prayer. Four years older than me, he was in high school already, so only something significantly wrong could explain his presence there.

"I'm sorry to interrupt, ma'am," he said to the teacher. "But I gotta take my brother with me to the hospital. It's our mom."

I was crying by the time we walked out the middle school's front door. Blaming myself, for wanting a way to get out of class—and here came God like the asshole that he was, giving me an afternoon off at the cost of my mother's life.

"Stop crying, stupid," he said, all sadness gone from his face. "Mom's fine."

His shitty old car was double-parked, hazards on. We got inside. He asked me, "Where do you want to go?"

My breath hitched, at the enormity of what was being offered. "Anywhere?"

"Anywhere."

"Albany," I said. "The mall."

We went. We got ice cream. He bought me video games. We rolled our windows down all the way and the wind screamed through his car and we played our music too loud and the world was ours and nothing could stop us.

I'd remember that day often. *He's still in there somewhere*, I'd tell myself. *The boy who loved me, who wanted me to have fun, and fuck the rules if they tried to stop us.* But I'd remember it too on the day my mom got her diagnosis, when Colby was nowhere in sight, and on the day of her funeral, when he was likewise absent. I told myself it was his fault, his doing, belated karma for that afternoon of fraudulent freedom.

**INEXPENSIVE, MEDICALLY IDENTICAL SYNTHETIC BLOOD HAS EXTENDED THE AVERAGE** American lifespan by four years. We did that.

Bizarre solar energy capture modules, which resemble bushes heavy with tiny metal leaves, have slashed carbon emissions in the developed world by 40 percent, and every year it drops another 6 or 7 percent. We did that.

Floating city construction.

New opioids that aren't addictive.

Recreational pharmaceuticals to painlessly reach a thousand different states of mind.

Government offices that make decisions impassionedly, with no window for human corruption to intervene.

You were right, that I was good at asking questions. Like, how can we make cheap durable stackable homes?

Fungitecture was our first big success. A software-engineered mushroom species that grows fast and fills a mold to produce a material as strong as cement, as light as Styrofoam, and cheaper than cotton. Like shipping containers, but at a tenth of the cost—and they float. That's the most important part these days. Two percent of the planet's population live in fungitecture homes today, including all thirteen thousand citizens of Tuvalu, floating over the ruins of their sunken homeland, and the eighty thousand climate refugees they've admitted.

I had so many questions, once I let myself start asking them.

**"AUSTIN?" YOU SAID, SEEING MY FACE GO SLACK. "ARE YOU OKAY?"**

I nodded. I wasn't. That random damaged man had moved on, but I had not.

"I'll think about it," I said.

You nodded. Your eyes locked onto mine and didn't let me look away.

Once you'd gone, I opened up a sex app. Growled or woofed at a couple dozen gentlemen in the space of two minutes. Started typing more detailed messages to the ones who'd growled back.

It was no good. My heart wasn't in it. To my great shame, something other than sex was weighing too heavily on my mind.

I dialed the number Colby had called me from. He answered immediately, his voice a whisper:

"Austin?"

"Hey, brother," I said, hating how my heart hurt at the weakness in his voice, making my own more cheerful in allergic response. "This your new cell number?"

"It's a strip mall pay phone," he mumbled. "Across the highway from the motel they kicked me out of."

"Jesus Christ, Colby," I said, looking at my phone for the time at which he'd called me. "You mean to tell me you've been standing by the pay phone for the past three hours?"

"Sitting, actually. The sun is nice."

"Florida's got that going for it, at least."

"Florida's not so bad," he said. "I don't know why you guys are so down on it."

Doors slammed, somewhere. On his end, or mine? Tears somehow blurred my ears as well as my eyes. Made me unsure where and when and what I was. "You okay, Colby?"

"Yeah," he said, which is what he always said, no matter how badly he was bleeding. "Sorry about before. I just kind of lost it."

"Of course you did, buddy. You almost fucking died."

"Dying didn't scare me," he said.

"What did?" I said, and now it was me who was whispering. "What happened, Colby?"

**I'D GONE TO VISIT HIM ONLY ONCE.**

The fights, I didn't mind. They weren't my thing, but there was something pure about them, something compelling in the contest of brains and brawn. What gay boy is completely immune to the spectacle of two sweaty mostly naked men grappling on the ground together?

What made me sick were the managers and promoters who made or broke a fighter. The moneymen behind the scenes who kept good fighters grinding away in obscurity while less skilled ones got the big matches that made the big money. Shit was unfair, what the fuck else was new?

So I stayed away. From all of that, and from the rest of his messy life. The drugs, the girls, the alcohol. When he called me, high, I'd talk to him. Listen, mostly. Big theories about the deeper meaning of a Pixies song or a Biggie lyric. Elaborate plans for how to work the system, game the gamers, get the stardom he deserved. *I know how they work, how they think,* he'd say, *how to play them, how to burn it all down.* He refused to see how screwed he was, how helpless.

That optimism, that faith—it had always been as alien to me as his addictions. And just as pitiable.

Then, though—on the phone—I envied his belief so acutely that it actually made me cry.

**I SAID GOODBYE. I HUNG UP THE PHONE. I BOOKED THE TICKET TO TALLAHASSEE.**

**YOU SCARED ME, CYBIL. OF COURSE YOU KNOW THAT. THAT WAS ALWAYS YOUR THING,** a look and attitude you cultivated. Spiky hair, jagged metal jewelry. The whole way to the airport, I kept thinking about you. A good sign, I thought. It showed I was less scared for my brother's life.

I had to take three buses to get to the airport. Construction of the Trillion-Dollar Fool-Proof Flood Locks had shut down most of the FDR, and the Real Estate Riots were in their sixth month. Not being underground, I could research you at greater length.

Intensity: that's what I saw when I looked in your eyes. That's what most people saw, to judge by all your press coverage. Now, though—I don't think that's what it was.

**I READ ELEVEN OF YOUR INTERVIEWS, IN THE INTERMINABLE TRIP BETWEEN LA GUARDIA** terminals. You called out the tech sector's casual solipsism, skin-deep liberalism, how we talked a good talk but never walked the walk, how we were "woke" on Twitter and thought that was enough. How the world was going to hell in a handbasket, and would continue to do so until the Silicon Valley rank-and-file—and workers everywhere—got off their fucking asses.

And somewhere in all of that, something else started to shift.

You quoted some old novel or short story: *It's not enough to hold justice in our hearts like a secret. Justice must be spoken. Must be embodied.*

**SOMETHING ELSE YOU UPENDED. THE CASUAL CYNICISM OF OUR POLITICS; THE TOXIC** partisanship; the hypocrisy and the apathy. Now we look back at those dark years and shiver, seeing clearly how close we came to becoming another one of history's cautionary tales of a nation committing suicide via nationalism.

•••

airport.

"I'll tell you later," he said. "I just ... I needed to hear your voice. It's stupid. I'm sorry. I know you're real busy these days—"

"I'm on my way," I said. "I'm at La Guardia. My flight leaves at five fifteen."

"Really?" he said, and I heard the tears start on his end, which kicked them off on mine as well.

"Of course," I said. "Of course I'm coming, Colby."

HERE IS WHAT HAPPENED. COLBY DIDN'T TELL ME UNTIL WE WERE TOGETHER, LATE that night, eating bad delivery food on the bed of a better hotel than the one he'd almost died in.

Try to picture the scene again, Cybil. The half-dead junky on the bed; the rain in the open door. The two cops. Flashing squad car lights. Christmas music from the next room over. *Rudolph the Red-Nosed Reindeer* too loud on the television; citizens of the Island of Misfit Toys singing songs of their lost homelands.

Lying on his side, mid fetal position, Colby locked eyes with the cops and begged. Waited for them to reach for the cuffs.

"The Black cop, I think maybe he recognized me," Colby said. "Maybe he saw me fight, somewhere. Maybe that's why he did what he did."

I guess that's possible. It's also possible that my brother wanted desperately to construct a narrative where something other than blind dumb luck kept him from spending the rest of his life in prison.

Whether he was a fight fan or he just didn't want to do the paperwork or he really and truly had the Christmas spirit—or all of the above—or for some other unknowable reason altogether, the Black cop put his hand on his partner's shoulder and tilted his head at the door, and the other guy looked startled, nodded, and then followed him out.

The strobing lights darkened. The car started. The cops drove away. Christmas carols continued, coming from across the wall. Colby cried and did not move, not for a long long time. Convinced they might come back. That they were fucking with him. That Florida could not contain such good fortune, not for him, anyway.

"IT'S AUSTIN," I SAID, WHEN YOU ANSWERED THE PHONE. IT WAS 2:00 A.M. IN Tallahassee—*What time is it in New York?* I thought to wonder, but of course it was 2:00 a.m. for you too. My brother slept in the bed beside me. My hand was on his arm, like I could hold him tight and keep him safe, like I could save him from a savage world. "I'm in, Cybil. If you'll have me, I want to be your partner."

ONE WEEK LATER, I WENT HOME ALONE. HOPEFUL, BUT NOT CONFIDENT. COLBY IN rehab. His stuff in storage. His word he'd call every day. His unuttered plea to come live with me when he got out.

You called me for lunch, this time at Little Poland. A booth in the back. Plates and plates of pierogies. Our second meeting, and I didn't know 'til it was over that I had been on the clock since I'd sat down, that I already worked for you.

"What's wrong?" you'd asked. "When we met before—and even today— you're troubled by something."

So I told you. Just the basics: big brother's an addict; almost died; didn't.

"Bring him in," you said. "He needs a job, or something. Right? If he's gonna stay sober. We can find something for him."

OF COURSE THERE'S MORE TO COLBY'S RECOVERY THAN ME GETTING ON A PLANE. Most of it, you already know—most of it was your doing—how you fig- ured out, over Tasty Hand-Pulled Noodles, where his skills sets were: how effortlessly he could play people, the same way he'd played my middle school math teacher; how he, like you, wanted to *burn it all down*. How you bought him a suit and set him up with three investor meetings. It's a long complicated messy story, but it isn't this one. This is just the story of how I said yes to him, and yes to you. I'm telling it to you, Cybil, but I'm also tell- ing it to myself. I'm finally understanding: we can change our minds, and we can change the world, but we can't change who we are.

FOR YEARS I WONDERED: WHY ME? HOW DID YOU WALK AWAY FROM THAT BENCH— and from me, with all my arrogant disinterest—and think, that's the one? Even when I called to tell you I was in, I felt sure I'd flunked the interview. I knew why I wanted to be on your team, but I couldn't imagine why *you* wanted *me* there.

I get it, now. I know why you chose me. Telling this story, writing this progress-report-turned-obituary, I see what it was. And I can turn this obit- uary into a goodbye letter.

Those other hotshot Titans of Iterative Modeling Software, boys and girls juiced up on the *WIRED* star treatment and exploding social media mentions—any one of them would have been smarter, more ambitious, more energetic than me. You always saw people as they were, no matter how they tried to hide the truth of who they were from others and from themselves. You'd have seen the hunger in their eyes. The hustle.

What you saw in my eyes was pain. Panic. The certainty that I'd lost my brother forever—the knowledge that maybe it hadn't happened that day, maybe he'd survived *that* overdose, but he wouldn't survive the next one, and if something didn't happen to disrupt the status quo of his long, lonely self-destruction, he wouldn't stop taking the risks that would eventually put him on the wrong end of a back-alley switchblade or a tainted shot or a fatal shared syringe.

You saw it in my eyes, and you recognized it.

### YOU, TOO, ARE A CITIZEN OF THE NATION OF THE SICK.

I see that now. I never did before. I bought your PR, even though I wrote most of it and so I should have known it was bullshit. That haunted look that sometimes flickered in your eyes—I imagined it was intensity, drive, the inexhaustible determination to bring a better world into being against all the formidable forces that opposed us.

And, yes, you contained all those things.

But you were something else as well. You were sick. Sick enough to one day take yourself out of the world completely. None of us noticed. Your fans, your followers, your henchpeople ... we never saw your sickness. That's our failing, our guilt to carry—that we saw you every day, and never saw you. Never offered you the help you needed.

### I'M SURE YOU'RE GONE FOR GOOD. I KNOW BETTER THAN ANYONE HOW YOU THINK.

You chose a death that would leave no body. There are millions of them out there. You chose something mysterious and open-ended enough that the vultures who've tried for years to take us down would never have a corpse to point to, to know when it's time to pounce. But the bottom line, I believe, is: you chose death.

Colby is convinced he'll find you. As our charming, gregarious corporate strategist travels the globe, helping build worker-owned coop incubators and collectives of gutter-punk coders who analyze and refine and write manuals for the bizarre new fruits our tree bears, he scours every crowded street he comes down. Every market square. He even had the Nunnery craft an algorithm to scan for your face in the feeds of every public traffic cam on the planet. He loves you, the poor dumb puppy. Same as me. You made him what he is—although he thinks it was you *and* me, but I know how that's bullshit, how you were making me at the same time as you made him. You turned me into someone who believes in something, and you turned him into someone whose unconventional skill set could be put to positive use—and he can't accept that he'll never get to thank you for it.

AGAINST PROTOCOL AND ALL PROBABILITY, A COP TAKES PITY ON A SAD-SACK, BROKEN-down, dope-sick ex-fighter. A direct flight to Tallahassee is a few extra hours away, so an asshole boy takes a meeting he'd otherwise skip. A homeless man happens to walk by; happens to slow down to make eye contact; happens to smile. If any one of those things hadn't happened, my life suddenly looks a whole lot grimmer. So does the whole fucking world, for all I know.

Citizenship in the Nation of the Sick means knowing how fragile our happiness is, how accidental our comfort, how little it takes to turn the warmest sunny day into dark cold night. We citizens of the Nation of the Sick, we know that all we have is the people we love. The ones we can save, and the ones we can't.

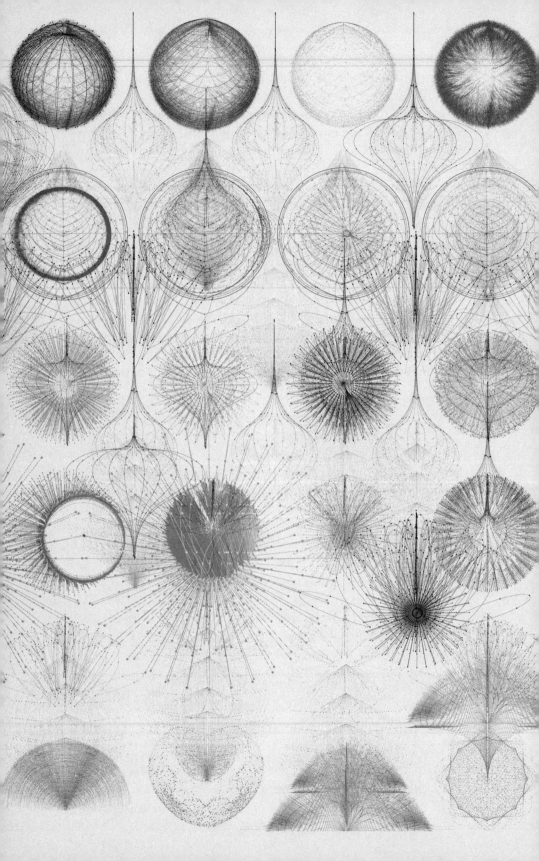

## Suzanne Palmer

RILEY WAS THE SMARTEST KID IN THE HIGH SCHOOL, WITH HER OWN SMALL CLIQUE of friends around her that seemed closed to all outsiders, so Jake did not expect the tall, gangly girl to nudge him with her shoulder on her way past him at the lockers. "You going to sign up for one of the after-school cleanup shifts in the classroom science lab, for the extra credit?" she asked. "You could use it."

Science was currently his worst subject, and Jake was fairly sure he could spend every day sweeping and putting tablets back in their charging nooks and curling up cables and still not pass, but this was *Riley* asking, and he didn't think she'd ever even spoken to him before and probably wouldn't again, so he nodded. "I guess, yeah," he said.

She nodded. "Cool-oh," she said, and continued on as if she'd never even paused, the blue fuzzy deeley boppers she'd glued onto her minder bobbing out of sight as her posse closed back in around her and they melted into the rest of the students milling about in the halls.

"You have three minutes twelve seconds until the start of your next class," Jake's minder said, in its low, perfectly unemotional, perfectly horrible voice. "It takes you on average two minutes fifty-eight seconds to traverse the distance from your locker to the classroom, so you should begin now."

Jake slammed his locker shut. He wanted to take his time, arrive late just to spite the thing plugged into the side of his head, but it would dutifully report even the briefest tardiness to his parents, and that just wasn't worth it. At least math was next; he usually came out of that remembering the entire class.

He made it in the door and to his seat with six seconds to spare. Ms. Lang was already at the smartboard, swiping across it from the virtual dock to bring up pages and images from where they'd left off the day before. Above the board, the classroom monitoring light was a steady green. "Okay, everyone," she said. "Back to quadratic equations we go, and get a little extra practice in before Friday's quiz."

There was a brief, familiar sting at his temple, and the world lurched forward a few seconds. There were suddenly two more windows open on the board, and the teacher had hopped a step to her left and was turning. Looking around the room at other puzzled faces, Jake wasn't the only one who'd gotten zapped. The kids without minders were snickering, and Ana was laughing into her hand. Jake's eyes met those of several other classmates as they all focused on Danny's desk, which now had a red light on it. "What?" Danny protested. "I don't even know what I said!" His own minder had gotten him, too.

"You said—," Ana started, and then there was another jump, and the teacher was facing them, cross now, and pointing at Ana with her board wand.

"Don't start this, or we'll never get through this lesson," Ms. Lang said. "I remind folks to watch their language and pay attention. If the entire class passes the quiz and your group average is a C or better, I'll bring in cookies next Monday. Okay? Now, back to the board . . ."

For once, the rest of the class went without interruption.

English was even more of a blur than usual, because Maya Angelou was apparently full of things his parents didn't think he should hear, and unlike random conversation, which was keyword-based, the minder could match a flagged portion of text as it was being read aloud and zap the whole sequence. He hated the feeling of having lost nearly half an hour of his time and of knowing that all his classmates with the orange-covered Good Parenting Code Redacted Edition of the book were just as robbed and confused as he was.

Last period was study hall, and he slunk down to the open lounge and sank into a chair, staring at nothing for a while.

Some kids, when they hit eighteen, chose to keep their minders, or went back to them not long after. He'd once seen an elderly man, way too old to have ever had one in childhood, with the familiar black half-hemisphere clamped onto the side of his head over his thinning white hair. Jake's father was always telling him he had to keep his as long as he lived under their roof, and his mom always stepped in to say of course he would, he was a *good boy* and *raised properly*.

Being glum about it all took up about half of study hall; when the room teacher started giving him looks, he took out his orange-covered history book and got started on his homework for tomorrow. Some texts removed unapproved material seamlessly, reworking sentences so the absence was invisible, but this one was just a reprint of the original with black blocks where the offending text once was, and the longer the block the more he despaired about passing the class. Foolishly he'd thought Ancient Rome would be a pretty safe subject, going in.

The end-of-school bell rang. He got up and packed up his stuff in his backpack and was being carried along with the great tide of students toward the buses when he remembered Riley. "Minder," he said, "tell my mom I'll be staying after school to do some volunteer cleanup for extra credit."

"This has now been communicated," the minder said a few moments later. "Dinner tonight is spaghetti Alfredo, your favorite, so don't tarry."

"Yeah, thanks," he said. He turned and made his way down to the science wing, past the ninth-grade poster projects and into his classroom. Riley was there, leaning on a desk playing idly with the blackboard remote, saving and closing windows one by one. Her tight-knit crew was all there. Kitt, a short, blond girl with cat ears on her minder, was taking washed lab gear from the dish drainer beside the sink at the back of the room and sorting them one by one into the equipment drawers. Nate, who was one of the few black kids in the school and got ten times as much crap as the other nerds, was pushing a variety of spitballs and dust around the floor in a meandering path toward the center with an old yellow broom, his minder sporting a painted-on Jolly Roger that had come and gone a couple of times, no doubt as part of a battle of wills with his folks that sometimes, Jake had noticed, ended in bruises. Erin, whose minder had been coated in purple glitter, was carefully lowering the window blinds trying to make them perfectly even. And Jonathan, a large, happy kid with disheveled light brown hair who had no minder at all, was making a giant stack of metal disks and blocks and magnets.

"Jake," Riley said.

"Hi," he said. "Where's Ms. Scott?"

"She trusts us to get through the checklist and not cause trouble," Riley said. "We've been doing this all year for her. She'll come by after the late bell and lock up."

"Okay," Jake said. "What do you want me to do?"

"Come help me with the glassware?" Kitt said. "Unless you're clumsy."

"Not too much," he said. He shrugged off his backpack, dumped it on a stool, and went back to the sink. "You guys do this every week?"

"Tuesdays and Thursdays," Nate said. "We all need the extra credit to fill in gaps, you know?"

"Not me," Jonathan said, as his pile of magnets fell over. "My brain is my own, and my grades are, too. I'd just rather hang out with these losers than have to go home."

Ever since freshman year, the five of them—Riley, Jonathan, Kitt, Erin, and Nate—had been best friends, a perfectly self-contained, self-sufficient social circle. The oddness of them inviting him here struck him again, and he lost track of which drawer he'd last returned pipettes to. "Why me, though?" he asked.

"A few months ago, Harris and Deke were bugging my little sister Lynne in the lunchroom, calling her fat and four-eyes and spaz," Erin said. "You stepped in and threatened to beat their asses if they ever even spoke to her again."

"Oh yeah." Jake hadn't known the two of them were related, though now that he thought about it, the resemblance between the two dark-haired girls was obvious. "Harris and Deke are jerks. Anyone would have done the same."

"No," Erin said. "No one had, all year, until you."

Jake shrugged. "I'm big. Everyone thinks that must mean I throw a hard punch and like to fight. Doesn't matter that I've never hit someone in my life." He almost had, once, and that had been close enough.

"Yeah, well, I'm glad you were there," Erin said.

Behind them, Nate got up and shut the classroom door, and leaned against it.

"What . . . ?" Jake started to ask, but Riley put a finger to her lips, then pointed at Jonathan.

Jonathan scooped up a handful of the magnets he'd been playing with, and threw them one at a time to the others. Then he walked up, a big grin on his face, and slapped Jake on the side of the head.

"Ow!" Jake said, taken by surprise. "What, you want to test me to see if I'll actually hit someone? Because now I'm game."

Jonathan stepped back, and held up his hands in surrender. "Look around, my man," he said, "look around and be free."

Jake curled his hand, trying to decide if he should hit this kid, when he caught Riley's eyes. She pointed at her own head—no, to her minder, and the magnet now stuck right onto the side of it.

The other three had done the same.

"We've been keeping an eye on you since Lynne told Erin about the cafeteria," Riley said. Jonathan began humming the *Twilight Zone* refrain.

154

"You don't go out of your way, but you're pretty smart, too, and it's your fucking minder that keeps you from being a top student."

"We shouldn't be having this conversation. You know our minders record—" Jake stopped midsentence. Riley had just dropped a swear in his presence, and he'd heard it, and *remembered* it. There was no skip, no awful blink as his minder took the moment away. Carefully, he reached up and touched the magnet.

Nate laughed. "Ooooh, you should see your face," he said, then more seriously, "You caught on fast."

Jake's eyes immediately went to the classroom monitor, whose light was off. *Right*, he thought. *School hours are over.*

Jonathan whacked him hard on the shoulder. "Welcome to the club, Jake-o."

"Uh . . . okay," Jake said, thinking it through. "The magnet scrambles the electronics, keeps it from working. Keeps it from recording?" Otherwise it'd be a short-lived freedom at best, and it wouldn't have a happy ending.

"Yeah," Erin said. "There are other things that disrupt the minders, so a little bit of static and missing timestamp on the recording doesn't stand out. As long as no one says anything, or does the magnet trick anywhere outside this room and this group, we're pretty safe. This is our secret, just us. Not even Lynne knows."

"You won't say anything, right, Jake?" Riley asked. "Not to your parents, your brothers and sisters, other kids here in school? And not your friends?"

"No," Jake answered. "I'm an only kid, and I don't have any friends. Not really." *Not anymore.*

"Well, you do now," Kitt said. "Just don't let us down. We're taking a risk on you."

"Why, though? I mean, this is cool and all, but it's a lot of risk just so we can have a Swearing in a Classroom Club."

"Because one of us can get through the checklist with time to spare. Five—*six* of us gives us a lot of extra time," Kitt said. "Jonathan?"

Jonathan reached for his own backpack, and pulled out all the class textbooks, not a single one of them with an orange cover. "You get tested on the full coursework, even the parts you can't hear because your mommy and daddy don't want you learning about communism and atheists and art boobies," Jonathan said. "If you're heavily minded, you're lucky if you can pull a just-passing grade average out of working your ass as hard as you can being perfect at everything else. Maybe you don't even graduate. This isn't a swearing club, it's Study Club. We take turns doing the cleaning and catching up on whatever reading we need to do, thanks to my lovely collection of uncensored textbooks."

"And what's in it for you?" Jake asked.

Jonathan dropped down onto a stool and thumped the top of the stack of books with his fist. "You all help me study, too, because I'm not as good at some of this stuff. Especially math. You good at math, Jake?"

"Math, yeah," he said. "Just don't ask me what the last month of science has been about."

Jonathan slung a book across the lab table at him. "Welcome to the sordid underworld of climate science, my man."

RILEY'S CREW, JAKE REFLECTED AS HE WALKED FROM THE BUS STOP BACK TOWARD his house, had things pretty solidly figured out. Since the minders could only record or respond to audio, normally they'd only "static up" for a few minutes at a time, when they needed to discuss things. Otherwise they took turns studying and cleaning, the cleaners making small talk and a constant soundtrack of noises. "In movies," Kitt had told him, "the people who make all the background noises for a scene are called Foley artists."

The school's American history teacher was named Mr. Foley, and he hated any noises at all in his classroom while he was lecturing—he once threw a smart pen at a girl for chewing gum too loudly—and when Jake pointed this out, they'd all laughed.

It was like he suddenly had friends again, out of the blue, and the knowledge that the minder could be circumvented seemed trivial in comparison.

When he got home, his mother greeted him at the door with her usual hug and kiss, then unlocked the minder and gently tugged it free from his head. "You've got a bit of chafing and redness around your implant again," she said. "You're just growing too fast. We'll get you a larger model next fall, okay?"

"Sure. Thanks, Mom," he said, hanging his backpack up on the coatrack behind the door as she plugged his minder into its dock to charge and upload its day's recordings to his parents' cloudpod. In a brief flash of daring, he added. "Maybe I could just not wear it tomorrow, to give my head a chance to heal?"

His mother shook her head. "You've got a perfect attendance record. I'd hate to see you miss a day if you don't need to."

"I wouldn't have to stay home," he said.

She *hmmmm'd*, an affectionate sound, and gave him a sad half-smile. "I know school must seem like a good and safe place, but that's because your minder keeps you away from the bad language, the wrongheaded ideas meant to confuse and lead you astray. If only we could afford private school for you, like Connor's parents ..."

"It's okay, Mom," Jake said, quickly. He didn't want to talk about his former best friend. Ever. "I like school, even with my minder. It's just hard sometimes, feeling like I'm missing things and I don't even know what they are."

"Poison," his father said from the kitchen doorway. "You're missing poison. Be grateful and go wash up for dinner."

BY THE THIRD STUDY CLUB HE HAD THE ROUTINE DOWN, INCLUDING RILEY AND Jonathan's system of hand signals, and there was very little chatting necessary, which also meant less static on the minder's recording to raise suspicion. During one of the blackouts, he handed off Jonathan's history text to Nate. "I might actually ace this test, for once," he said.

"Be careful about that," Riley cautioned. "If you do too well, or there's a sudden jump in your grades, your parents might start wondering why."

"I hadn't thought of that," Jake admitted. The idea that he might have to deliberately flub some of the questions bothered him immensely. But how much should he actually know about things like the Tulsa Massacre, completely absent from his own orange-bound, censored text?

"You'll get used to it," Kitt said. "You just gotta graduate or get to eighteen, whichever comes first."

"I don't think it's gonna be that easy," he grumbled. "Not if my dad has any say in it."

"Yeah, well, my dad talks with a cane," Riley said, and the others all looked down at their books, uncomfortable and quiet, until she dramatically groaned and got up from her chair. "My turn to sort the stupid beakers," she said. "Your turn with social studies, Erin."

IT WAS ODD NOT DREADING BEING AT SCHOOL EVERY DAY, BUT GETTING OUT OF the house in the morning had become torturous as he worried about not seeming too happy, for fear his parents would note the change and become suspicious.

Proof he'd overacted his misery came when his mother stopped him on his way out the door a few days before the holiday break. "You miss

Connor, don't you?" she said. "You two used to be best of friends until he went to Angel Valley."

"I don't miss him at all," he said, more forcefully than he intended, and his mother frowned at him.

"I was thinking I could invite his family over for dinner, maybe over break," she continued. "Would you like that?"

"They won't come," Jake said, grabbing his bag, as she pushed the minder down onto his head and it clicked into place. "Don't you get that? We're not good enough for them anymore, because I still go to the public school and even though I have a minder I'm not one of them, and I won't ever be good or pure enough anymore. So thanks, I get to be looked down on by both the Angel Valley kids and the regular kids because instead of being brainwashed the old-fashioned, socially acceptable, rich kid way like precious Connor, I've got this Frankenstein thing stuck in my head deciding what I'm allowed to think."

"Jake!" His mother was shocked and visibly upset.

"I'm gonna be late," he said, and slammed the door behind him before he could hear her counterargument, or worse, crying. He'd get more than enough yelling when he got home anyway.

**"YOU OKAY?" KITT ASKED WHEN HE CAME INTO THE SCIENCE LAB JUST AFTER THE** last bell and threw himself down at one of the lab benches. She slid a magnet to him across its hard black top.

"Parents," he said, after he put it on. He looked around the room, Erin and Riley both leaning over the same history text as Jonathan doodled chickens all over the smartboard. "Where's Nate?"

"Volunteered to vacuum the library," Erin said. "Ms. Carrie says they don't pay her enough to monitor what kids are reading once school hours are over, as long as they have a legit reason to be there. I think he cares more about reading for fun than actually studying. He's been working his way through *The Two Towers* since sometime last spring."

"The what?" Jake asked.

Jonathan whistled. "You haven't heard of it? At *all*? Man, you have a sad-ass life."

Jake stood up so quickly he knocked a beaker off the lab bench, and it shattered all over the floor. "You wanna say that to me again?" he demanded.

Riley got between them. "Hang on, hang on. We're on the same side here. Jonathan, you want us to call you out on stuff you got no choice in?"

"I didn't mean anything. It was a joke, okay?" Jonathan said.

"Say sorry," Riley said.

Jonathan rolled his eyes and let out a deep, melodramatic sigh. *"Sorry, Jake,"* he said.

Riley turned around and whacked Jake on the shoulder. "Say you're good now," she said.

"Am I?" Jake asked.

"You better off without us?" Riley asked.

"No," Jake said. He sat down, then immediately stood up again to get the broom. He didn't meet anyone's eyes as he swept up the broken glass into a small pile, then an even smaller one, before finally swatting it into the dustpan. "Don't say that ever again, Jonathan, and then I guess we'll be good."

"I really didn't mean anything," Jonathan said again, more sincerely this time, and Jake sat back down and picked up the math textbook.

He slung it over onto Jonathan's desk, where it landed with a loud thump and nearly slid off the far side before Jonathan slapped his hand down on it. "Chapter fourteen," Jake said. "Log, Sine, and Cosine. Try the first ten exercises and then if you still don't get it, I'll try to explain."

TWO WEEKS LATER, JAKE BURST INTO THE SCIENCE LAB, WAVING HIS GRADE SLIP. "I got an A!" he shouted.

No one leapt up to high-five him. "What subject?" Kitt asked warily, handing him his magnet.

"Science," Jake said, feeling a sudden pang of anxiety deep in his gut as he put the magnet on. "The astronomy unit. Can't be anything—"

"Did you get the questions right where you have to use the ruler to measure how far away the galaxies are on the picture and plug them into the formula to get their real distance?" Nate asked.

"Yeah," Jake said.

"And how far away were they?" Nate asked.

"Like fifteen thousand light years—"

"So how long did it take that light to get here?" Nate asked.

"And how old do your parents believe the universe is?" Riley added.

Jake stared down at the slip of paper.

"If they don't go looking at the test questions, you're probably okay. If they do, and you haven't already used it, try the 'lucky guessing' excuse," Erin said. "That won't work too many times, though."

"Oh," Jake said, and couldn't help but think, *but I got an A*.

DINNER WAS QUIET. TOO QUIET, EVEN THOUGH JAKE HAD NOT BROUGHT UP HIS GRADE. He ate his chicken and peas, indecision roiling in his mind. Was it suspicious that he didn't mention it, and did that make him look guilty? Or was bringing it up only drawing attention to something that might otherwise go unnoticed?

Later, as he was sitting at his desk in his room doing his homework, he could hear his parents talking in the front hall. "The diagnostics say it's okay," his mother was saying. "He'd tell us if it didn't seem to be working right."

"Would he?" his father asked. "Remember what he said to you the other morning? This is supposed to protect him, not turn him defiant."

"All young men are defiant," his mother said. "You were. Oh, were you a troublemaker when I first met you. You're lucky I believe in second chances."

"I was, and that's how I know I was wrong. Maybe we should look into our finances again, see if we can scrape together enough to transfer him to Angel Valley after all."

There was silence for a few minutes, and the muffled beep of the minder station as it finished charging his unit. He heard his mother let out a long sigh. "I don't think Angel Valley would be a good place for Jake," she said at last. "I don't like how exclusionary those families become."

"The Ducettes sent their boy there. He and Jake were best friends, and we get together with them regularly."

"We used to," his mother corrected.

"We've all been busy—"

"I invited them over for dinner. Abigail laughed at me, told me not to bother them again, and hung up."

"What?" his father sounded genuinely surprised. "I'll talk to Robert after church this Sunday, find out what that was about. There must just be some misunderstanding."

"Sure," his mother said. "Let me know how that goes. In the meantime, let's not talk about Angel Valley. Not yet."

"ANYONE ELSE SEE THE NEWS LAST NIGHT?" NATE ASKED.

Riley snorted. "Like any of us are allowed to watch the news. Since when are you?"

"Since my mom fell asleep on the couch with the remote unlocked," Nate said. He tapped the side of his minder, next to the magnet. "Did you know these are now illegal in sixteen states? North Carolina just outlawed them too. In seven it's considered child abuse."

"Too bad we don't live in North Carolina," Erin said. "You know that's never going to happen here."

"Maybe if enough states do it, it'll become a national law," Jake said. He remembered the right word. "You know, federal."

"Not soon enough for us, though," Kitt said.

"We could get lucky and it could be soon," Nate added.

"You think so?" Riley asked. "You know they don't even have any good studies on the long-term effects of frying kids' hippocampuses—"

"Hippowhat?" Jonathan asked.

"Part of your fucking brain," Riley said, and Jake found himself flinching reflexively at the swear, even though it couldn't currently trigger his subverted minder. "That's how it works, right? You've got a little wire stuck right into your brain so this piece-of-shit tech can disrupt your damned memories and keep things it doesn't like from getting converted from short to long term, as if modifying our very thoughts inside our heads somehow keeps us purer than accidentally remembering a few random syllables."

"I thought it just made it so you don't hear it," Erin said.

"It's worse. You hear it perfectly fine, and then it takes that away from you. Like someone breaking into your room and stealing your favorite thing but as soon as it's out of the room you can't even remember you ever had it to start with, only you just know something is missing you can't identify," Riley said. "It *should* be criminal, except who gets to decide that? The goddamned thieves themselves."

"You okay, Ri?" Jonathan asked.

"No!" she said. "How can I be? How can any of us be? It's not enough to get around their mind control. We should be fighting it."

"And how do we do that?" Nate asked. "We don't have any power."

"My father would ship me off to Angel Valley," Jake said. "There's no *getting around* that. That would be the end. I don't want to lose what freedom I have, even if it's not much."

"Yeah, well, you can't go off to Angel Valley anyhow," Erin said. "My sister Lynne has a crush on you for saving her from the cafeteria bullies. You're her personal hero."

Riley stood up and threw her magnet down on a lab bench. It landed with a sharp bang like that of a gunshot. Everyone went dead silent. "This is all just a waste of time," she said, and left the classroom.

Other than the door banging closed again behind her, the silence lasted long after she was gone.

It was Nate who broke it, with a small cough. "Things are really bad in her house," he said. "I mean, really bad even compared to the rest of us. At least I think for most of us we can say our parents are trying to do what they think is best for us, even if we don't agree. Riley's folks—"

"It's about absolute control," Kitt added. "She doesn't even have a door on her bedroom."

Jake whistled. "Okay, that definitely sucks," he said.

Erin closed her textbook and got up, grabbing the bench wipes. "They don't like that she's smart," she said. "It's a threat. So she has to pretend she's not."

"She's smarter than any of us," Jonathan said. "But her folks talk about sending her off to one of the church work farms when she flunks out of high school. You know, one of those places surrounded by barbed wire? Ain't there to protect the corn."

"Shit," Jake said, the swear weird in his own voice, powerful and shameful.

He glanced up at the clock to cover his sudden embarrassment. "Oh hey, time's almost up," he said. He double-checked that he'd made it through the section of the science chapter blacked out in his own textbook, then closed the unabridged copy and handed it back to Jonathan.

Together they finished up the last of the lab cleanup and reluctantly left their magnets behind.

While he was walking down the hall toward the school's front entrance, Jake's minder beeped. "Jake, your mother has requested that you stop at the corner store on your way home to pick up some green leaf lettuce and a small carton of half and half. Do you wish to hear these items repeated?"

"No," Jake said.

"Do you wish your agreement to be communicated to your mother?"

"Yes," Jake answered. His minder hadn't spoken to him in several days, and it was a jolt to be reminded that his brief escapes from it were only that.

He was still lost in thought about Riley and the minders and hippocampuses when he turned down an aisle in the corner mart and nearly collided with someone coming the other way. "Connor," he said.

"Jake," Connor said, taking a half-step back as if even being near him was distasteful. He was wearing his Angel Valley school uniform.

The last conversation they'd had, Jake had strongly considered punching him. He'd grown though, right? He had other friends now, better ones. He could be the bigger person. "How've you been?" he asked.

"On the higher path," Connor answered. "We're the future of the new America. There will be no room for the tainted trash of humanity then, and I have no time for you now."

Whatever Jake said in reply, his minder stole the moment from him, but there was at least some satisfaction in seeing the suddenly pale kid hurrying away as if he'd been bitten.

**163**

He regretted, just a tiny bit more, not punching him. Either way it was worth whatever trouble he was going to be in when he got home and his minder reported the encounter.

JAKE'S MOM CAME INTO HIS ROOM JUST AS HE WAS ABOUT TO TURN OUT HIS LIGHT. She sat on the edge of his bed and patted his knee. "You know you were wrong to use that word?" she asked.

"I'm not even sure what word I said, but yes," Jake said. "I was wrong. But he used to be my very best friend and we did everything together and he called me *trash*."

"Your father and I expect you to make better decisions," she said. "But we know you have a good heart, and as wrong as you were to speak to anyone that way, Connor was also in the wrong for what he said to you. We want you to be the best Jake you can be, and if you are, only God can judge you."

"I'm sorry, Mom," Jake said.

"I know," she answered, and kissed him on the head before turning off his light. She left his door cracked partway open, and as he lay there overwhelmed by guilt and lingering anger, his father's shouting still ringing in his ears from earlier, she reappeared in the doorway.

"That's a word you shouldn't know, and I trust that you won't ever speak it again," she said, "but if you absolutely had to use it, you chose the right occasion."

That gentle understanding pushed the remainder of the anger out of his system and left him trying to find sleep while wallowing in his own shame and regret.

•••

the door of the science lab, and handed him his magnet. "Where you been?"

"Grounded for two weeks," he said.

"For the A?" Nate asked.

"No, for calling someone a bad word I can't remember," Jake answered. "But knowing how angry I was when I said it, I'm pretty sure the grounding was justified."

"I was worried you got caught and sent off to Angel Valley," Erin said. "Then we'd be stuck trying to teach Jonathan math."

"Lucky for you I'm still here," he said. He had been unsure whether he should come back, but that uncertainty disappeared.

"Angel Valley isn't the worst," Riley spoke up from where she was sitting on a lab bench with a science book on her lap. "You know they're working on glasses, to also censor what we can see?"

"No one would—," Erin started to say.

"My parents would, if it were available," Riley said. "And minders aren't just used to censor out science, or things that disagree with faith—*any* faith, or even all faith. You know there are some where you can't hear anything at all spoken by people of a certain gender? Or race?"

"What?" Nate said. "That's not cool."

"I'm just saying, for every person out there working on a way to get around, or get rid of, minders, there's someone working on making them worse. And some of the things they're trying ... Angel Valley wouldn't be so terrible in comparison."

"How do you know all this?" Kitt asked.

"I have connections," Riley said. "You all might be looking for a way through, but I'm looking for a way out."

"What happens when you find it?" Kitt asked.

"Then I'm gone."

"And what about us?" Erin asked.

"You gotta make your own choices," Riley said. "I—Oh *shit*."

She was staring at the classroom door. Everyone else followed her gaze, to find that Erin's sister was standing there in the open doorway. "Lynne!" Erin shouted.

"I thought maybe I could help," Lynne said. She was blushing, and studiously not looking at Jake.

"How long have you been standing there?" Erin asked, pulling open a drawer to find another magnet.

"A few minutes," Lynne said. "Since you were talking about some kind of new glasses."

Erin had found an extra magnet, and ran toward her alarmed sister. "It's too late," Riley said. "Her minder heard everything. Get her out in the hall. Plan B."

Erin nodded, a jerky, panicked gesture, and pulled Lynne out of the classroom.

As soon as the door shut behind them, Jake turned to the others, his heart thudding in his chest. "Plan B?"

"Plan A was if one of us was caught. Plan B is all of us," Kitt said. "As soon as her minder gets downloaded, the shit is going to hit the fan. They'll know we've circumvented the minders, if not how."

"Can we just, somehow, erase the record in Lynne's minder?" Nate asked.

"Not without pulling it off her head, which without properly unlocking and disconnecting it could cause brain damage," Riley said. "Trust me, if there was a way, I'd have found it by now."

Erin stuck her head back in the door. "I'm going to take Lynne for milkshakes on the way home. That will buy us an extra hour or so. Good luck everyone."

"Are you safe," Riley asked Jake, "if you go home?"

"I think so," he answered. Safe with his mother, sure, but his father . . . "I'm not sure."

"If you're not safe, get to Jonathan's house. That's our safe point. Don't tell anyone where you are going."

Jonathan nodded. "The only danger at my place is my mom's cooking," he said.

"And then what?"

"Like I said earlier, everyone has to make their own choices. I didn't think it was going to be anything more than rhetorical this soon, though," Riley said. "And Jake, if you don't come? It was good knowing you. Try not to let anyone steal who you are, or who you want to become, away from you."

JAKE LEFT THE SCHOOL WITH HIS MAGNET STILL ON HIS MINDER, HIS HOODIE PULLED up over it so no one would spot it; part of him wanted to broadcast to the world how to get around having your brain under external control, but Riley had convinced him that as soon as the hack was publicly known, the minder manufacturers would find a way to prevent it from working. Better to pass it on, whisper by whisper, friend to friend.

He started walking toward home almost as many times as he started walking toward Jonathan's house, and ended up mostly doing irregular loops around town, until he found himself near the elementary school playground and took a seat on one of the swings. The sun had set and the sky was shifting quickly toward night when he took his magnet off.

"You are late for dinner and should immediately return home. It will take you twenty-one minutes to walk from this location," his minder immediately said. "You have fourteen messages from your mother and six from your father. Would you like to hear them?"

"No," Jake said. "Send a message to my father and give him my current location—"

"Your current location has already been provided to your parents," his minder said.

*Of course*, Jake thought. "Tell him to meet me here, just him, and if he calls the police or brings anyone else along he will never see me again. You got that?"

"The message has been sent," his minder said.

"Good," Jake said. "Now shut up." He put his magnet back on, then relocated from the playground across the street to a small park. Feeling both foolish and paranoid, he sat on the ground behind a bench, where he could see through the slats but not be easily seen.

His father arrived twenty-nine minutes later and stood by the slide, looking around.

It took Jake a few minutes of watching his father to rally his courage; then he walked toward him, hoodie still up, hands in his pockets to hide their shaking. His father spotted him as he crossed the street and stood straight and unmoving until they were face to face.

"Jacob," his father said.

"Dad," he answered.

They regarded each other for several minutes, before his father spoke again. "I don't expect I need to tell you how angry and disappointed I am," he said.

"No, I think I got that," Jake said. "How's Mom doing?"

"She's hurt and scared. What do you think?" his father snapped. "She made me agree to come out here and meet you on your terms despite that. So this is your play: what now?"

*What now?* Jake asked himself. He had no idea what his "play" was, except to be honest.

"You know, Connor used to be my best friend. I didn't need other friends, because we had done everything together since kindergarten. Then his parents sent him off to Angel Valley, and not only is he no longer my friend, he's no longer *Connor*," Jake said. "And I just want to keep being Jake, because I think I'm a good kid, but I can't know who I am and what I am while my entire reality is being decided by someone else. Even if that's you."

"The job of a parent is to protect—," his father started to say.

"The job of a parent is to guide their children," Jake interrupted. He'd never interrupted his father before in his life, and didn't know how he dared to do so now. "This minder isn't guidance, it's locking me in a tiny little box. And maybe, yeah, there are some pretty rotten things out there in the world, but this feels like you're assuming just knowing about them would make me want them, and that's not fair. I can learn about something without accepting it unconditionally, I can talk it through with you to understand why you have different opinions, and I can form my own judgment, but I can't know if I agree or disagree with anything if I can't know about it at *all*."

"Why me? Why aren't you throwing this sales pitch at your mother?" his father asked.

"Because Mom would get it. I don't know if you can. And as much as I love you, if you can't understand then I can't come home," Jake said. "I'm giving *you* the chance to think and decide for yourself, because if I'm going to demand that right, I owe it to you."

His father blew out a long breath through his nose, and stared at him long enough that Jake felt like his knees were going to give out and pitch him onto the ground. "I don't like this," he said at last. "I don't like that you defied me, that you've been sneaking around behind my back to do something you explicitly know is forbidden, and that somehow you think a few words can make that okay."

"I don't need you to like it, and I don't need you to stop being angry, but I need you to respect me," Jake said. "I need a second chance as much as I need to offer one to you."

"Wait here," his father said, and abruptly turned and walked away.

It was almost two hours later when his mother appeared in the park, walking toward him shivering as he sat at the top of slide. "Jake?" she called out.

167

He climbed down, and she wrapped him in a tight hug, and for a moment he was afraid she wouldn't let go. Then she stepped back and held him at arm's length for a moment, before she reached into her coat pocket and pressed something into his hand.

The minder key.

"There are some other kids missing. Friends of yours, I guess," she said. "Do you know where they are?"

If they were still at Jonathan's house, they'd be gone soon. Riley and Nate, certainly; he didn't know about the others, and didn't know where they would go. "No," he said.

"You demand your so-called freedom, but it comes at a cost," she said.

"What?" he asked.

"Don't ever lie or hide things from us again. And don't ..." Her voice faltered.

"Don't what, Mom?"

"Don't stray from the path of being good," she said. "Not for the sake of trying to be smart, and certainly not just because you feel a need to disagree with your father. Because if you think your minder was a pain, you have no idea how many arguments you have set yourself up for instead."

"I will do my best," he said.

"Then come home. Your dinner is ice cold by now."

"I don't mind," he said.

RILEY, NATE, AND KITT WERE GONE. KITT CAME BACK ABOUT A MONTH LATER, NOT wearing her minder. She was no longer allowed to speak to him or the others remaining from the Study Club, but whenever they saw each other in the halls they made a beeline to intersect and high-five as they passed.

Jonathan was very relieved that Jake was still willing to help him with math.

Lynne seemed to blame him somehow for what happened, and his hero status was gone, but every once in a while she'd dump her tray next to his in the cafeteria and sit with him. He assumed that meant either she'd eventually stop being mad at him and knew it, or she was strategically re-upping their association to keep Harris and Deke off her back.

His mother had not lied when she told him he'd set himself up for a world of arguments with his father; every single thing that the minder or his old censored text would have kept from him, his father had to hash out in excruciating detail. It was exhausting and felt pointless, except as the weeks and months wore on, his father's arguments became less a belligerent lecture and more a heated exchange of near equals.

He missed Riley and Nate, and worried about them, wondering if they'd gone their separate ways or had stuck together, wishing he knew where they were or even just that they were okay. There was no one he could ask who would know, but wherever they were, he knew they were joining the fight against minder tech.

On the last day of school before summer, he was cleaning out his locker and found a dog-eared paperback book called *The Fellowship of the Ring* inside, a note sticking out from the middle.

*For when you still need to escape*, the note said, and was signed *Nate*.

At the bottom of the page was a postscript. *PS*, it read, *This is like a year overdue. Please return it to the library when you're done and pay my fine. Thanks.*

Jake tucked it in his bag, shut his empty locker, and went home.

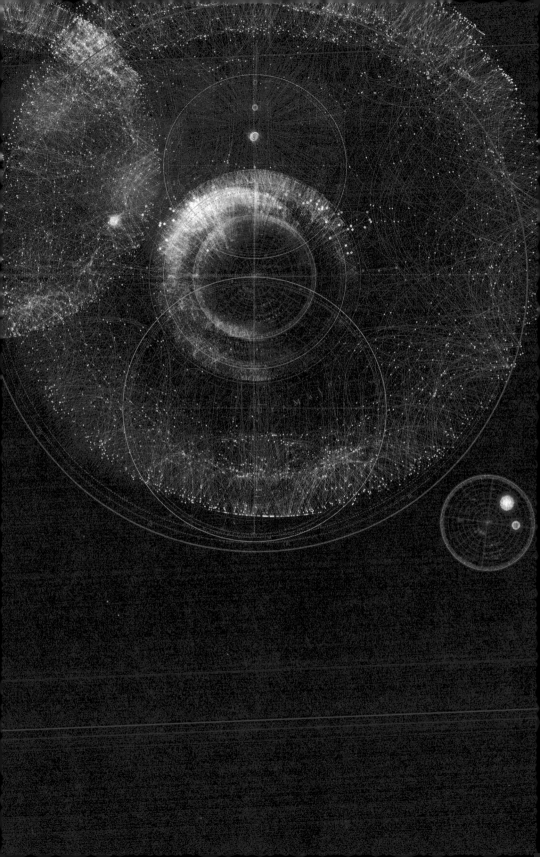

**Annalee Newitz**

EDWINA WAS SMEARING BACTERIA ON THE WALLS OF THE PRESCHOOL LUNCHROOM. Bubbles snapped inside the rubbery microbial slime. Toddlers love to lick walls, and the more they came into contact with this stuff, the better. Stepping back to survey her work, she stripped off her compostable gloves and set the timer on her watch. In twenty minutes, this coat would be dry and she could apply the next layer of cultures. The San Francisco Health Department had just released an updated childhood microbiome recipe, including dozens of newly sequenced bacteria that contributed to intestinal health. Every public building for children had to buy the stuff. That's where Edwina's nonprofit came in. AFDC, or Aid for Facilities with Dependent Children, raised money to infect schools and day cares that couldn't afford it.

As it set, the bacterial slime slowly became translucent. Edwina eyed it critically, looking for spots she'd missed. Her watch pinged her contacts with a chat message, and she blinked open a text window that hovered to the left of her vision.

Chester: I got tickets to the vocaloid show with TrixxieStixx! Are you in?
Edwinner: Hell yes!!!!!
Chester: Meet at DNA Lounge at 9?
Edwinner: Sure. See u there!

She blinked out, feeling the now-familiar twinge of guilt. Chester was adventurous and adorable—lovable, even—but so was the other person she was dating. Augie, the bookworm who got social anxiety in crowded places, could spend hours talking to her about the minutiae of public policy and never get bored. Chester and Augie were both smart and great in bed. She genuinely liked them, too. And it wasn't as though Edwina had ever lied to them about what was going on. They knew about each other, had even met a few times, and neither of them seemed particularly concerned. But after almost a year of dating the two of them, Edwina was starting to feel like she had to make a choice. If she was honest with herself, she was falling in love with Chester and Augie, and that didn't seem right. Especially not if she wanted to get married and have kids in the next decade.

Sighing, she put on a new pair of gloves, opened the next tub of bug-laced goop, and plunged her hand all the way inside. It felt like her fingers were trapped in an amorphous, unnameable embrace.

FOUR DAYS LATER, EDWINA MET UP WITH DAISY AND ALYX FOR THEIR WEEKLY BAD rosé night at whatever wine bar had the lowest rating on Eater. When she arrived, Daisy was gushing about a new product from ProTox called Eternalove.

"It's this amazing breakthrough based on vole hormones!" Daisy enthused. "ProTox says it will turn anyone completely monogamous! Isn't that insane? Apparently there are slutty voles and monogamous voles, and they figured out the hormone that switches monogamy on and off. So when you're ready to get serious, you and your sweetums take Eternalove and never want to fuck anybody else again!"

"Sounds real," Edwina snorted, rolling her eyes. "I mean, you know this is the company that said they'd genetically modified wheatgrass to cure cancer."

Alyx smirked and raked their fingers through their long bangs to create a perfect fan over one eye. It made them look like one of those 1990s Filipino movie stars that Edwina's aunties loved. "The placebo effect is real, man," they said. "Scientists have measured it! Who cares if it's a sugar pill or some vole sweat? I bet this will make some people super happy. And it's going to make a great subplot on *Natural Urges*."

Alyx and Daisy ran a startup that made video content for brands, and their biggest success was *Natural Urges*, a streaming series sponsored by ProTox. Mostly it was about the trashfire romantic lives of twentysomethings, who always happened to be using ProTox's latest beauty and health gadgets. This season, Daisy played an elementary school teacher who was secretly addicted to hookup apps.

Daisy continued. "Yeah, we're going to have my character dose her new boyfriend with it, but then he falls in love with this other guy and it's going to be really hot." She rubbed her hands together gleefully.

"Well, that does sound pretty cute," Edwina admitted. "I'm glad ProTox is still supporting queer stories. I guess that means the show can't officially air in North Carolina anymore."

Alyx and Daisy shot each other pained looks. Alyx fiddled with their glass and sighed. "ProTox might ask us to make a special hetero-only version for North Carolina markets," they said. "We're not sure. I mean, people there could use a streaming proxy to watch the original version. But ProTox thinks that more states will pass Family First laws, and we'll need alternate content for them."

The three of them got into a heated conversation about the Family First legislation in North Carolina, the first state in the United States to ban homosexual and transgender content. There were a ton of protests, and some hollow threats from streaming companies like Disney that didn't want to deal with the nightmare of figuring out what to block and how. But in the end, the law had passed. Now Disney was scrambling to perfect its geolocation algorithm so that nobody in North Carolina would be exposed to gay wedding videos or transgender superheroes or whatever else they decided was dangerous.

"I can't wait for Disney's 'whites only' channel," Alyx muttered, draining their glass.

"Ugh. I honestly wouldn't be surprised." Daisy lowered her eyes to look at her hands, pale pink next to the brown of Alyx's arms resting on the table.

It reminded Edwina of how she'd met Alyx back in high school. Alyx's family had just moved up to Stockton from Manila, and they were the first person at school to use nonbinary pronouns. Edwina thought it was badass, and kept looking for ways to catch Alyx's attention. Finally she overheard them telling a classmate that they missed banana ketchup, so she filled a little bottle with the sweet, vinegary glop from her mom's secret stash and brought it to lunch.

"I really hate this stuff," Alyx said, making a face. "It's just something I tell white people about when they ask me about the Philippines."

"That's a really random troll," Edwina said. "I don't get it."

Alyx cocked their head at Edwina, meeting her eyes for the first time. "People don't really want to know what it's like in other countries. All they care about are weird brands that aren't available here."

Edwina thought about the comment and realized it was true. The main thing her white friends knew about Filipino culture came from watching off-brand Filipino superhero shows on YouTube. But it was the "knockoff Captain Marvel" part that sucked them in, not the Filipino part.

Even back then, Alyx had a lot of thoughts about branding and social justice. Edwina vowed at that moment to be their partner in crime forever, and so far her plan was working out. She'd met Daisy much later, at a crappy job after college, but had the same feeling about her.

Looking at her two dejected friends, Edwina decided it was time to lighten the mood.

"So," she said. "I need relationship advice."

They both looked up with matching grins. "What's going on?" Daisy asked.

"Nothing bad. I mean, things are going really well with Chester and Augie, but I think I'm ready to get serious. I can't keep dating random people. I need to settle down. I want to start planning for kids and stuff."

Alyx looked alarmed. "You're going to have kids?"

"I mean, not right away. But in like five or ten years, yeah."

"Why do you need to settle down now for potential kids in ten years?" Daisy asked, pouring more rosé from the carafe.

"It seems weird to keep dating two people when I should be focusing on one."

"You could be polyamorous," Alyx suggested. "It's not illegal, even if nobody in North Carolina can watch shows about it."

Edwina thought about the polyamorous people she knew. They were all old white millennials who had a lot of drama on social platforms that nobody used anymore. When she thought about calling herself polyamorous, it felt wrong, like putting on a muumuu when she wanted to wear a tux. "I don't think that would work for me," she said finally. "It seems creepy. Plus I hate combining Greek and Latin words."

Alyx arched their right eyebrow ironically. "Good point. All sexual and gender identities should follow traditional rules of syntax."

"That's not what I meant," Edwina laughed. "I'm just saying it's not me."

Suddenly Daisy started bouncing in her chair. "Oh shit!" she yelped. "You should try Eternalove! We could do a whole story arc about this!"

"This is my actual life, Daisy. Not a branding campaign."

Alyx shrugged. "That's quite an on-brand thing for you to say, Miss Nonprofit Worker."

"Dude, I'm actually feeling kind of upset about this, and I don't want you to turn it into social commentary."

"Whatever. Stop debating," Daisy said, waving her hands to make them both shut up. "You should try it. Why not? ProTox sent us all these testimonials from people who swear it worked."

Edwina found herself taking the idea half-seriously. What harm could it do? Eternalove was completely unscientific, but as Alyx said, the placebo effect was real. Maybe it would be like flipping a coin, where merely the act of doing it would help her decide. And hell, she supposed it was remotely possible that the scammy ProTox gang could have licensed an elixir that actually did what they claimed. Twenty years ago, nobody would have believed that smearing germs on the walls of schools could save a whole generation from asthma and irritable bowel syndrome. Vole hormones might be the next game changer, like birth control pills in the twentieth century.

Shaking her head, Edwina realized she'd talked herself into it.

Digging through her enormous tote bag, Daisy finally pulled out a flat, rectangular box in matte crimson. It was embossed with silver lettering that read: "When the feeling is exclusive. Eternalove."

"This is a full course," Daisy said, handing it to Edwina. "On the house."

Alyx snorted with laughter as Edwina slid the box open and stared at the silver blister pack resting on a bed of cotton. Each pill was a tiny red heart. They looked like Valentine's Day candies.

"I'm going to regret this, aren't I?" Edwina sighed. "How do they work?"

"ProTox says you take them every day for a month, and you'll stop wanting to have sex with anybody except the person you love."

Edwina didn't think that's how vole cognition worked, but she'd already given up on plausibility at this point.

"It's a love potion, Edwina! It's magic." Alyx used the voice they used to pitch branding ideas to companies. "ProTox is about making fairy tales... *real*." They widened their eyes and did some jazz hands for full effect. Alyx was a master at wrapping sincerity up with irony, then enclosing the whole package in sincerity again.

"Okay, I'm doing it." Edwina popped the first pill out and put it on her tongue. She held it against the roof of her mouth until the sugar coating melted and she tasted its bitter core, imagining what it would be like to say goodbye to Chester or Augie forever. Then she gulped down half a glass of wine.

having some effect. Augie, still asleep on her futon, had never looked more beautiful. Her pale eyelids were lacy with faint purple veins and the sunlight turned her messy blue hair into a fiery sapphire. Edwina pulled the blankets away from Augie's right shoulder, kissing its plump curve, then biting into it lightly.

Augie opened one eye. "Hi, sexy." She rolled onto her back and fixed both green eyes on Edwina. "Would you say I'm more of an organic farm-to-table kind of meat, or something grown in a vat?" Augie worked in food safety, and a big part of her job involved inspecting synthetic meat facilities.

Edwina pretended to consider. "I'll need to taste you again."

Augie pulled the covers down further. "I wouldn't want to mislead consumers, so you'd better be certain."

As she licked Augie's stomach, Edwina felt her heart race and knew this was right. Completely right. The pills had worked their magic. "Augie," she whispered. "I think I'm in love with you."

Augie wrapped her arms around Edwina's head, pressing her belly hard against Edwina's cheek, awkward and intense. "Oh sweetie. You know I love you, too."

They spent the day cuddling, ordering Mexican dim sum from an indie delivery service app and bingeing on episodes of *Fae Killers*. The whole weekend was like that—quiet and romantic. Of course the pills weren't the cause. Edwina knew that. But as she'd suspected, taking them had helped her figure out what she really wanted. She and Augie took a long walk together on Sunday afternoon, climbing to the barren top of Twin Peaks, where the Army Corps of Engineers had planted a plaque in the ground 150 years ago. The lettering stamped on it had been erased by weathering, leaving only a smooth metal disc in the middle of a rocky outcrop. A few other brave people had hiked up here, too, rewarded for their perseverance by a sunset that filled the Bay with a deep rosy light. Wind turbines stood like giant robots over the water, churning invisibly behind their bird-proof screens, and fat crows made a racket in the air overhead. Edwina looked north toward the green rectangle of Golden Gate Park, obscured in part by the red spines of Sutro Tower, and took Augie's hand. They kissed as the sunlight was demolished by the earth's rotation, and Edwina was almost certain she was the luckiest person in the city that night.

THE NEXT MORNING AT WORK, EDWINA SWALLOWED ANOTHER HEART-SHAPED PILL. Somehow she'd gotten superstitious about them and wanted to finish out the whole pack. Only three more weeks of placebos to go—and a very awkward conversation with Chester. She managed to shove him out of her mind until he texted her at the end of the day.

**Chester: Want to play the beta of the Captain Marvel AR game? My friend Long is showing it at the Hurricane Warehouse.**

Edwina blinked the chat window out of existence. She didn't want to deal with this right now. But then he texted again.

**Chester: It's supposed to be the best Captain Marvel since the one in the 2030s. She can team up with Spider Gwen! The effects are amazing. I might try to play Spider Ham. Also, we could get curry Japadogs after. What do you think?**

This was what she loved about Chester. He was always going on adventures, and his job as a VFX designer meant he knew where the weirdest things were happening. Edwina paused, wiped all her work out of the air, and tapped her temple to go offline. What had she been thinking? That there was something she loved about Chester? She put her head in her hands and sighed. Maybe if she actually saw him, the feelings she had for Augie would blow this attachment to Chester away. She tapped her contacts back on, feeling bittersweet but determined.

**Edwinner: Sure! Meet you at the Hurricane!**
**Chester: See you soon, cutie!**

Edwina had gotten her hopes up for nothing. The moment she saw Chester's spiky black hair and half-smile, she knew she was still smitten. He was wearing a sparkly yellow shirt that perfectly set off the dark brown of his skin, and she immediately wanted to kiss him. Squashing her feelings down, she focused on her incredible weekend with Augie. She remembered their sunset kiss and was sure Augie was the one.

More and more people crammed into the warehouse space, and they shoved every last piece of furniture against the walls so they could interact when the show started. They all downloaded a character sheet and started figuring out their skills. Edwina was going to be She-Hulk, which meant she had legal acumen and enhanced strength. As she put on headphones and set her contacts to work with the warehouse AR network, Edwina realized this was the kind of fun she could never have with Augie. Most of the time, Augie couldn't deal with crowds. She liked routine. There's no way she would ever have gotten into an interdimensional war with a bunch of strangers playing characters from twentieth-century comics. Still, Chester would never have stayed in bed all day talking about the latest E. coli infection disaster. Or gone on a nature hike to a lonely peak.

After they defeated Magneto and Loki, Chester led her up Natoma Street to a food cart serving Japadogs with mango curry and various other bioluminescent sauces. They ate sitting on the curb, far enough from the cart that the chef wouldn't hear their reviews.

Chester elbowed her lightly and wiped some mango blobs off his cheek. "What do you think?"

"Ummm... I'm afraid I'm going to say no to mango curry Japadogs for the rest of my life."

He laughed. "What? I thought She-Hulk liked to eat everything."

A feeling washed over her like a blast from one of Doctor Strange's magical semiphores. Edwina pulled Chester to her tightly, giving him a long, hard kiss on the mouth.

"Hey, cutie, what was that for?" He looked simultaneously quizzical and horny.

The words tumbled out. "I think I'm in love with you." Once she'd said it, she knew it was true.

Chester's eyes widened. "Really? Because I think I'm in love with you too! You are absolutely, 100 percent, the greatest person ever." And he hugged her again, kissing every part of her face right there on the street, only a few yards away from the worst Japadogs known to humanity. It felt right, just like kissing Augie did.

The next morning, Chester made Edwina a perfect cappuccino with a rocket-shaped machine that he'd stolen from the set of a long-ago canceled video series. Edwina popped another pill out of the blister pack, swallowing it with a mouthful of foam. Taking these things had only made her life worse. Now she was officially in love with two people, which was the exact opposite of what was supposed to happen.

"Are you going out with Augie tonight?" Chester asked.

At the sound of Augie's name, Edwina almost spilled her coffee. She stared at Chester, unable to say anything.

Chester continued, seemingly unperturbed. "I was only asking because it seems like you two usually hang out on Tuesday nights. Isn't her day off on Wednesday?"

Maybe Edwina had told Chester that at some point. Or maybe Augie had told Chester, one of those times when they'd met. Finally Edwina found her voice. "I'm not sure what I'm doing tonight." She heaved a sigh. "I don't know what I'm doing at all."

Chester shrugged, oblivious to her internal conflict. "Okay, well let me know. We could go to that new Godzilla popup theme park in Oakland if you're free."

Why was Chester so casual about this? Now that they had used the word *love* to describe their feelings, shouldn't he be discouraging her from seeing Augie? And shouldn't Augie have been telling her not to see Chester? Edwina stared morosely out the window of Chester's flat, which had a panoramic view of the bay if you stood in the exact right spot to see between two other apartment buildings. Then she texted Alyx.

Edwinner: I think the pills are doing something weird to me, and now I'm stuck in this super awkward situation where I have no idea what to do.
Alyx777: I warned you.
Edwinner: What do you mean?
Alyx777: I said the pills were magic! Or maybe garbage? Either way, the kids are buying them like crazy. Did you see this week's episode?
Edwinner: I'm serious, Alyx. do you have time to meet up for happy hour?
Alyx777: Yeah. Sorry about the love trauma, boo. Let's meet at El Rio.

Edwina blinked away the words to see Chester smiling at her the same way he had last night, when he called her the greatest person ever. It made her want to cry, or have sex with him, or maybe both. They walked to the bus together, holding hands and talking about how Godzilla was still the best giant monster. It felt natural and unnatural at the same time. She couldn't wait to focus on something completely normal, like spraying bacterial precursors on birthing tables in a maternity ward to encourage the growth of microbes that used to live in people's vaginas.

ALYX WAS DRINKING A LONG ISLAND ICED TEA WHEN EDWINA ARRIVED AFTER WORK. Her hands still felt pruny from wearing gloves all day.

"You look kind of sick." Alyx sounded concerned. "Are you still actually taking those pills?"

Edwina settled on the bar stool next to Alyx and nodded. "I mean, they're just placebos or something. But I do feel strange."

"They might really have hormones. Or—I dunno. ProTox has some bizarre subcontracts with people who claim to do sorcery and shit. Who knows what's in there."

Edwina ordered a scotch and sipped it slowly. "I think the problem is that I need to break up with one of the people I'm dating."

"Why? Is one of them better than the other?"

"No. The opposite!" Edwina said vehemently. "It reminds me of how I always have to order a scoop of chocolate and a scoop of salted caramel when we go to Bi Rite. They are so perfectly, equally good that I can't imagine ordering only one. Does that make sense?"

"I think you're saying that Bi Rite has done such a good job of marketing their flavors that they've suckered you into always ordering two scoops instead of one," Alyx said. Then they saw the look on Edwina's face and reached out a hand to squeeze hers. "Sorry to be so snarky. I guess I don't get why you are so freaked out. My mom had seven kids, and she loves all of us. Like for real. Imagine if you had to keep track of what seven people like for breakfast, or what they want for their birthdays. She always has, though. If she can love seven maniacs, why can't you love two flavors?"

Edwina could feel tears in her eyes, and her contacts started to drift off her irises with an annoying string of error messages. She blinked them back into place and used one finger to draw circles on the bar with a blob of water. "I want to have kids. Nobody will let you marry two people and have kids with them."

Alyx looked more serious than she had ever seen them. "You know that's bullshit, right? I can't think of a better place to raise kids than with grown-ups who love each other." They drummed their fingers on the bar and seemed lost in thought for a moment. "Marriage is like every other brand that has staying power. Think about YouTube. It used to be part of a private company, and it was full of really bad stuff, like Nazis and crazies talking about rounding up gay people. But then YouTube spun off and became part of the public broadcasting network, and now it's all educational programs and people gardening and stuff. That was a major rebrand, but it worked. Most people don't even know that it used to be dangerous for kids to go there."

"And this is related to my situation how?" Edwina drained her glass.

"Marriage is another changing brand. It used to be only for cis heterosexuals, but now gay people can get married—at least, in a lot of places. People don't think of marriage the same way anymore. Even in North

Carolina, where they have those Family First laws, people are protesting. Here in California, you can create an indie brand marriage. And you know what happens to indie brands, right?" Alyx winked. "They get appropriated by giant megabrands. Pretty soon, ProTox will be marketing a placebo for people who want to fall in love with more than one person. I guarantee it."

Edwina shook her head. "I don't know."

A text from Augie was blinking in the corner of her vision.

**Augustales: Hey are we hanging out tonight, or are you with Chester?**

What the hell was it with both her sweeties being perfectly fine with an arrangement that she was taking pseudoscientific vole hormones to shut down?

**Edwinner: I'm actually with Alyx. Can I come over in an hour or so?**
**Augustales: Sure. I'm just cleaning up cat barf and reading the latest caloric intake report from the FDA lol**

Alyx had finished their drink and was wearing their usual sardonic expression again. "Off to do some rebranding?" they asked.

Augie's apartment was in one of the old luxury condo blocks built in the Castro around the turn of the millennium. It had been divided and subdivided until all the original three-bedroom, two-bathroom layouts were studios, most with shared bathrooms. Somehow, Augie had scored an en suite bathroom, which still had most of the fancy fixtures from the 2010s. Even the heated floor worked.

Flopping on the futon next to Augie, already in her frayed plaid pajamas, Edwina screwed up her courage. "We should talk about our relationship, I guess."

"Yeah, I was thinking that, too. I mean, this weekend was intense. But I'm glad." Augie leaned over and kissed Edwina's ear.

"I really want to get married and have kids one day. I mean, not now, but not super far into the future either."

"Me too. Same."

"And I think Chester does, too. I mean, I need to talk to him more, but I'm pretty sure he does." That wasn't exactly a great way to put it. But, as Alyx said, she was making her own indie brand from scratch. The first try is always a little rough.

Augie was nodding. "Yeah, I like Chester. He has good taste in women." She laughed and ruffled Edwina's hair. "We should probably hang out more often if we're going to have kids together, though."

Edwina felt wobbly. "We probably...should? I mean, you're not mad?"

"If I were mad, I would have stopped seeing you like eleven months ago when you started dating him. Look, I need to have lots of quiet alone time, and you like to go out and eat cockroaches while dancing to electronic flute music. I'm glad you have somebody who can do that with you, because I definitely won't."

Edwina hugged Augie, smelling the detergent in her pj's and marveling at how soft the skin on her neck was. She wished she could put this emotion into bacterial slime and spread it everywhere so that everyone could feel it. "I love you," she whispered, knowing those words were inadequate. Still, she planned to say them to Augie about fifteen million more times before her life was over.

CHESTER GOT OFF WORK EARLY THE NEXT AFTERNOON AND VISITED EDWINA AT A school in Richmond that had never been infected before. AFDC had raised money for the school after the *San Francisco Wave* reported that kids in the neighborhood were getting bronchitis and Crohn's disease at extremely elevated rates. Volunteers were scrubbing down the walls and spraying everything with agar to promote bacterial growth. The kids had packed all their stuff into neon yellow plastic bags so that it wouldn't get soggy, and the teachers had taken down all the posters and art. She could see bright squares on the bulletin boards where pinned-up calendars had prevented the corkboard from fading. A few pencils stuck out of the high ceiling; apparently kids still shot them up there with rubber bands. Somebody had left a dead monitor folded up in a dusty corner. Pretty soon this would be a room full of chaos again, swirling with good bacteria.

Chester sat down in one of the tiny plastic chairs and looked at the bucket of slime Edwina had just cracked open. "Should I eat some of this? Will it help my digestion?" He hovered a finger over the goo, a quizzical expression on his face.

Knowing Chester, he probably would eat it if she told him it was okay. Which technically it was, though really it wasn't. She needed every ounce of it for the kids.

When the volunteers went on break, Edwina sat down on a pink beanbag next to Chester and stripped off her gloves. "So I was talking to Augie about our relationship."

"You mean your relationship with her or your relationship with me?" Chester asked.

"I guess . . . both?"

"Did you talk about how we are all going to grow old together and raise amazing kids who eat microbes with banana ketchup and understand the difference between Captain Marvel and Captain America?"

Edwina awkwardly scrunched the beanbag closer to Chester and leaned her head against his leg. "Were you guys texting each other or something?"

"No. I was just saying what I hoped you talked about. Especially the Captain Marvel part. You're not breaking up with me, are you?" Chester looked uncharacteristically anxious.

"I meant it when I said I love you. But I love Augie, too. Is that okay?"

"It's been okay the whole time we've been dating, cutie. I like Augie. We totally *should* have kids together. I mean, not me and her. But you know—me and you, you and her, all of us. One day." Chester shrugged. Edwina looked up at his crooked smile and thought about how he was always up for an adventure. Of course he was up for this.

She turned her attention back to the classroom, its walls drying, and thought that maybe Alyx was wrong about what this was. It wasn't like rebranding. It was like science. Her mother and grandmothers grew up with spray bottles full of antiseptics and antibacterials, killing everything in the environment until people started getting sick. It turned out that people needed all that bacteria swimming around in their guts and mouths and various other unmentionable tubes. The immune system is a million different moving, living creatures, and if kids aren't eating dirt then they are probably missing out. Before Edwina was born, people would have called this classroom dirty and sick. Now they knew it was the healthiest place for kids to grow up.

Edwina pulled the blister pack of Eternalove out of her pocket and stared at it. She didn't need magic to figure out her future. She already had evidence-based analysis. Kissing Chester on the knee, she stood up and crossed the room to the waste bins. Each candy heart made a satisfying snap as she popped it out of its plastic shell and into oblivion.

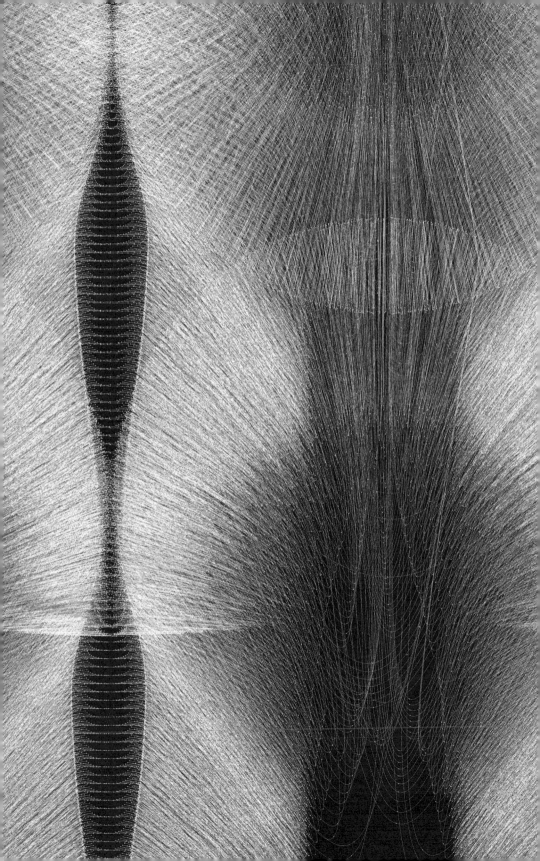

# 11 THE MONK OF LINGYIN TEMPLE

## Xia Jia

translated by Ken Liu

### Perfuming the Altar

OCTOBER. LINGYIN TEMPLE WAS SHROUDED IN AN AUSPICIOUS DHARMA CLOUD.

To the southwest of the temple ran a narrow footpath, named Tianzhu Path. Lined with tea gardens, ancient shrines, simple huts, and bamboo groves, it followed the course of a babbling brook.

It was evening, when there were few tourists. A lone man dressed in black walked along the path: hair white, face lined, a few deep furrows in the brow, as though etched permanently into the flesh. The rain had stopped so that the road and the surrounding vegetation all glinted wetly, and fragrant golden-red-yellow-white osmanthus petals covered the flagstones like stars. He walked deliberately, one step after another, as though lingering over the beautiful scenery, or perhaps slowed down by a troubled mind.

A woman dressed in white stood in the road up ahead. She pressed her hands together and bowed in greeting. "Householder Zhou, you may call me Xiao Wang. The Venerable Zhengxuan asked me to wait here to welcome you." She wore little makeup; her dark eyes were lively. A small scarlet red mole sat between her eyebrows, as though painted there in vermilion.

The man returned her greeting. "Thank you."

The two walked together down the path, side by side.

"Isn't the scenery breathtaking?" asked Xiao Wang.

The man nodded. "Indeed. I had no idea this path was here."

"Everyone's so used to riding the LINGcart now; so few take walks," Xiao Wang said. "Before, every time I visited, I'd take this path—it felt like a way to cultivate a connection to Buddhism. No matter how much my heart was troubled, a walk would always sort it out a bit."

"Do you often come to the temple for self-cultivation?"

She shook her head. "I can't call myself a believer. I'm just here to help with the Liberation Rite of Water and Land. Householder Zhou, did you come specifically for the Liberation Rite?"

He said nothing.

Xiao Wang went on. "The Liberation Rite will release from suffering all beings of land, water, and air in the six destinies. The deceased will be freed, and the living will accumulate merit. It's rumored that the Abbot of Lingyin Temple, the Venerable Zhengxuan, will be retiring this year. That is why this Liberation Rite is even more solemn and grand than usual, and the number of participants so numerous."

"The Venerable Zhengxuan is getting on in years, isn't he?" the man asked. "I believe he's been the abbot for a long time now."

"He's seventy now. He renounced the secular life and became a monk at Lingyin Temple eighteen years ago, and was elevated to the position of abbot eight years ago."

"Eighteen years . . . ," the man murmured.

"Eighteen years sounds like a long time. But looking back, it seems no more than a dream."

The man looked away and said nothing.

"Householder Zhou, do you know the Venerable Zhengxuan?" Xiao Wang asked.

He started as though awakened from a dream. "I was fortunate to meet him once many years ago. Back then . . ." He paused, and then added, "Fortune makes fools of us all."

Xiao Wang stopped. "We've arrived."

The two stood before the western gate to Lingyin Temple. The sun hung low over the horizon, and the sky was covered in golden-scarlet clouds. The screeching of birds returning to their nests for the night filled the woods.

"The Liberation Rite will begin tomorrow and last a full seven days and seven nights," Xiao Wang said. "The altars have been prepared, and the dharma cloud sima-boundary barrier is active."

"The sima-boundary barrier?"

"The rite is sacred. To prevent disturbances from nonparticipants, only those with proper identification may enter and leave. Please extend your right hand, palm forward, like this."

Following Xiao Wang's lead, the man held out his hand. He felt his palm connect with an invisible wall, light as cloud, cold as water, hard as diamond and glass. A golden glow, lotus-shaped, blossomed at the point where his palm made contact with the invisible wall, and spread apart like ripples in a pond. He looked up and watched the glowing light gradually fade into the sky. The entire Lingyin Temple complex was enclosed in this giant invisible dome.

Photographs of him and Xiao Wang flickered across the sima-boundary barrier. Following a clang like the striking of a bell, the photographs faded and dissolved into a white mist, revealing a circular opening.

He was surprised. LINGcloud was a new technology based on carbon nanocomponents. The nanocomponents could capture the water molecules in the air and drift about freely like a real cloud, changing colors and textures to bring about dreamlike experiences for users. Many were predicting that it had the potential to replace silicon-based electronics within the next decade. However, due to the high production cost, it was still limited in deployment. He had not expected Lingyin Temple to possess so much LINGcloud, or to have programmed it to achieve nearly magical effects. Apparently the many fanciful rumors about this place were at least partly based on truth.

The vermilion temple gates swung open, and the air filled with the harmony of monks chanting sutras.

Xiao Wang whispered, "They are about to start the ritual incensing and purification of the altars. Householder Zhou, please follow me."

After a moment of hesitation, he stepped over the threshold. The gates swung closed, shutting out the chittering birds.

## Repentance Ritual of the Emperor of Liang

INSIDE THE CRAMPED CHAMBER, A WOMAN SAT ALONE ON HER REED MACE PUTUAN cushion, chanting sutras.

Dressed in a monk's robe, she held a strand of prayer beads in her hands. Her long hair, unkempt and wild as weeds, draped to the floor.

The chamber was tiny, about three paces across from west to east, and also three paces across from north to south. There was only one bed, one table, one chair, one person, one putuan. Sunlight slanted in from a single window, dragging the lone shadow slowly across the floor.

She couldn't remember how long she had lived like this. Every day, she got up at the clacking of the temple clappers, before the sun had risen. She ate, prayed, studied, day after day, year after year. Long ago, before she had entered the temple, she had thought those who renounced the secular life whiled away their time in leisure, but the reality was anything but. The temple was like an intricate windup mechanism, and from morning to eve, every person, every stick of incense, every chant of *Amituofo* was precisely regulated, assigned its place. More than once she had wondered, who set up such a system? But no one told her, and all she could do was follow.

She had tried to rebel against the rules and was prepared to bear the consequences. But there was no punishment; no one rushed in to beat her or berate her, check her work. However, there was nothing to do inside the sealed chamber. She didn't eat, didn't drink, covered her head with the blanket, and slept until hunger bit at the inside of her belly like the mandibles of insects, forcing her to get up and eat. And when she had eaten her fill, what awaited her was boredom.

She had tried to occupy herself in various ways, but nothing lasted. She had explored every nook and cranny of the room, searching in vain for an escape route. The whole room was sealed by a sima-boundary barrier, and even mosquitoes couldn't find a seam. She had tried to smash the window with her chair, or to ram her head against a corner of the table, but always, the barrier recognized her intent and immobilized her with shocks before she could succeed. In despair, she had lain on the ground, hoping she would go mad or die. But she didn't die, and she didn't go insane. Her body, like the temple, held to its own patterns. When meals were brought to her, she slowly crawled over to eat. She slept until she could sleep no more and got up to read the scriptures. Evening drums followed morning bells; the stars rose as the sun set. Her hair grew longer and longer.

She learned to live according to the rigid timetable of the temple, turning herself into a component in the machine, revolving day after day with precision. She learned to sit in Chan meditation, to chant the sutras to pass the time. From an anxiety-ridden existence, she learned to find pockets of calmness, to achieve a state of oblivion as to her identity and condition. She learned to accept waves of assault by emotions such as anger, frustration, hatred, and bitterness, to let them pass through her and depart. She learned to take care of her body, to eat well, sleep well, get some exercise, and keep the chamber spotless and neat.

She requested a needle and thread so that she could repair the robe and sheets that she had torn in her despair. They acquiesced. Clumsily, she threaded the needle, thinking: *What if I use this needle to blind myself? Would I be able to leave then?* She covered her eyes and fumbled in the dark, but soon give up the thought.

After that, every afternoon they'd send some clothes to her room for her to mend. She treated the work as a reward: other than eating, sleeping, and chanting sutras, she had at last found something else to do. She became even more industrious, hoping that she could request even more: books other than Buddhist scriptures, pen and paper, cards and games, or even a meal with meat. Some of her requests were granted, but not others. Bit by bit, she tried to expand and enrich her life inside the chamber.

Clappers sounded outside her window.

She stopped chanting, opened her eyes, got up, and stretched her limbs. She had finished her work for the day. Before the evening meal, she was allotted half an hour of free time.

She held the prayer beads, palm up. A dharma cloud rose at her feet, slowly resolving into a miniature real-time projection of Lingyin Temple, every hall, shrine, tree, even blade of grass in vivid detail. This was the first day of the Liberation Rite of Water and Land, and the temple complex thronged with the faithful and their pious incense. The sound of monks chanting reverberated in the ancient halls. She waved her hand gently, and the projection enlarged like the time-lapse video of a bamboo grove in spring, until it was life-sized.

She found herself inside the Hall of the Medicine Buddha, where forty-eight venerable monks led a gathering of householders in chanting the *Repentance of the Emperor of Liang*, dating from the sixth century. It was said that Emperor Wu of Liang's consort, Empress Xi, was a cruel person in life. After her death, she turned into a giant snake and sought out Emperor Wu in a dream, begging for help. The emperor, a devout Buddhist, then found nine eminent master monks to compose a repentance text that would release her spirit from the lower realms and allow it to ascend to heaven. The text had been passed down through the ages and was said to be particularly effective in releasing the souls of sinners from suffering.

She didn't join in the chanting, but observed the throng of the faithful carefully, guessing why they had come. For what deeds were they repenting? Behind those seemingly kind and peaceful faces, what kind of sins and crimes lay concealed? She remembered the many she had known in her former life; though she chanted sutras and prayed all day, she never once sought relief for them. The dharma cloud projection was so realistic that she could even smell the incense, sense the heat from the skin of the faithful. Almost subconsciously, she reached out to touch the face of a young householder. Her fingers passed right through the projection, touching nothing but emptiness.

She lost interest and decided to go elsewhere. As she turned, however, her gaze swept over the face of an old man, his hair snow-white. Under those permanently scowling brows, his searing eyes locked with hers. Shocked, she clutched the prayer beads and waved her hand, covering her face with her flapping sleeve. The dharma cloud projection vanished. When she put down her sleeve, she was back in her tiny chamber.

Her legs shaking, she collapsed to the floor. The robe was plastered to her chest and back with cold sweat. No, it had to be a mistake! *That man* couldn't have seen her. But that face... she would recognize that face anywhere.

She held up her hands in lotus mudra, and a wisp of dharma cloud fell into her cupped palms, turning into the Chinese character 业. The character flared up like a blossoming fire, and flowing streams of light, red and blue, twisted and entwined inside the flame, growing and declining by turns, mesmerizing. At the bottom of the fire, a giant red swirl roiled and turned, like a malignant tumor, or the bleeding eye of a demon. Sweating profusely, she set down her hands, dismissing the projection.

*You reap what you sow. One cannot escape the wheels of karma.*

The clappers sounded outside the window. It was time for the evening meal.

## Feeding the Burning Mouths

HE WALKED THROUGH AVICI HELL.

Blood and gore covered the ground. The legs of hungry ghosts soaked in pools of blood, tendons swollen and clumps of hair poking up like pond scum. He was exhausted but had to continue on. The moment he stopped, fire seared the bottom of his feet; the moment he stumbled against a hungry ghost, it revived and went after his flesh. He had no choice but to fight off the ghosts with his teeth and nails: digging out their eyes, extracting their organs, sucking out their brains. When he had eaten his fill, he knelt to wash his hands in a pool of blood and caught his own reflection. He had turned into a hungry ghost himself.

Startled awake, he found his hair plastered against his temples with a hot sweat that had soaked his pillow.

Gradually, the ceiling of the temple guest room solidified in his field of view, reminding him of where he was. The silvery moon illuminated a small patch of floor before the window, and he could hear the chirping of autumnal insects outside. He raised his hands; they were clean, devoid of blood. He pressed his hands together, rubbing away the cold perspiration.

Draping a jacket over his shoulders, he strode into the yard. Two ginkgo trees stood in the moonlight, rustling. The ground was covered in the fan-shaped leaves. He walked around the yard, listening to the sound of crushed leaves. He thought of the insects and worms hidden under the fallen foliage, abruptly halted, and felt the flames against his feet.

He was thinking of last evening, when he had participated in the ritual to feed the burning mouths at the Hall of the Medicine Buddha. "Burning mouths" referred to ghosts plagued by hunger, and the Three-Master Yogic Ritual for Feeding the Burning Mouths was intended to bring them relief, to free them from their pain. The ritual lasted from evening until close to midnight, and for the duration participants had to observe a strict fast. He and the other participants had sat on the ground, enduring hunger and thirst, praying for all the dead immersed in the sea of suffering. But he himself could find no relief. As soon as he fell asleep, he was plunged into the same nightmares.

He held up his hands in lotus mudra, a tiny 𑁍 rising between his palms. After a moment of hesitation, he pressed his hands together and rubbed hard, as though trying to crush some secret between them.

He turned and found Xiao Wang standing under one of the ginkgo trees.

"Are you having trouble with the hard bed here in the temple guest room?" she asked.

He chuckled mirthlessly. "I've been suffering from insomnia for a long time."

"The Liberation Rite involves a great deal of work. If you don't get enough rest, I'm afraid you won't be able to endure it."

"But aren't you also up?"

"I've always gone to bed late and gotten up late. Though I'm supposed to follow the schedule here at the temple, I can't change my habit. If you can't sleep either, we might as well talk."

"All right."

They sat down on a pair of stone stools. The autumn night breeze felt chilly.

He asked her, "Do you come to the temple often?"

"Not at all. But, I'd say that fate has connected me with Lingyin Temple."

"How so?"

"This mole on my forehead wasn't there from birth. When I was little, my parents once brought me to the temple to offer incense. I saw that all the statues of the Buddha had a red mole on the forehead and found the look appealing, so when I got home, I took a red pen and did the same to myself. But after the paint had been wiped away, a red mole gradually grew in its place."

"That really is fate then."

"Today's Lingyin Temple is nothing like the temple I visited as a child."

"I think I know what you mean. I'd heard so many rumors about the temple and thought most of it fiction. But now that I've seen the place for myself, I'm beginning to believe."

"What sort of rumors?"

"They say that starting with the Venerable Zhengxuan, many other brilliant minds have taken their vows at the temple and become monks. Today, Lingyin Temple is filled with talent and has the brainpower to be ranked as a leading scientific research institution. Indeed, rumor has it that several major tech titans have visited Lingyin Temple to offer incense and were given valuable advice by the venerable monks. Moreover, the Chan-science symposium, held twice a year at the temple, is oversubscribed by those craving enlightenment with their futurism. Some even claim that several so-called black technologies in recent years were given their magical qualities due to the involvement of the monks from here."

Xiao Wang laughed. "'Black technologies!' That's absurd. But it is true that Lingyin Temple conducts scientific research and is connected with the larger tech scene. The temple has two specific divisions involved in these matters: Wenshu Institute is responsible for science and technology, while Puxian Institute is in charge of charitable activities. Wenshu Institute is mainly responsible for the temple's daily management, but also participates in the development of AI hardware and software deemed helpful for

Buddist cultivation, such as using LINGcloud to bring the entire temple complex into the AI age and creating the 业 system for calculating everyone's karma progress. Puxian Institute, on the other hand, is more like a charitable foundation operated by the temple. In addition to direct money and material aid to the poor, it also supports various long-term projects to improve the lives of the public: medicine, education, environment, food, energy, transportation, architecture, urban planning, data security, technology ethics, animal rights, and so on. Though Puxian Institute tries to maintain a low profile in the media, an endless stream of representatives from various organizations comes to the temple to apply for funding for their projects, especially during the Liberation Rite. Puxian Institute is unique in that it doesn't decide where to allocate resources based on traditional measures of return on investment, but on the amount of merit and benefit to karma generated by each project. I'm at the temple mainly as a consultant to help the institute evaluate these applications."

"I see," said the man. "A charitable foundation hands out money without expecting anything in return.... Where does Lingyin Temple get the money?"

"When the Venerable Zhengxuan took his vows, he donated all his assets to the temple. Morever, each year, the temple receives considerable donations from the faithful. Lingyin Temple has quite a sum at its disposal."

He sighed. "I've heard that the Venerable Zhengxuan has led a hard life. His eldest son was born with a rare congenital condition that current medical technology couldn't cure. In response, he formed a research foundation dedicated to the disease. Later, his wife died in an accident involving self-driving vehicles, and as a result, he devoted all his remaining resources to the development of next-generation transportation technology, believing that a global network of personal transportation tubes could replace dangerous moving vehicles. Everyone at the time thought he was babbling an impractical dream. But eight years later, the first city-scale LINGcart network launched and received much praise. Yet, just when the future for his investment seemed so bright, he left the secular life and become a monk. The news shocked the world, and most reacted with skepticism and puzzlement. But now, looking back, I see that perhaps everything had been arranged by fate. Maybe he really is the Buddha, reincarnated to save the world, and had to experience pain and suffering first."

The two sat in silence as the ginkgo leaves susurrated in the night breeze.

Xiao Wang broke the silence. "May I ask how you became a Buddhist?"

His brow furrowed, and only at length did he answer, "My family also suffered a terrible tragedy, and I sought spiritual relief. However, even after years of cultivation, I cannot find release from the pain."

Xiao Wang pressed her hands together and bowed. "I wish you the best in your search."

## Releasing Caged Animals for Merit

SHE SAT ALONE INSIDE THE CHAMBER, HOLDING THE STRAND OF PRAYER BEADS. A dharma cloud enveloped her, connected to all her sense-faculties: eye, ear, nose, tongue, body, mind, heart.

A miniature projection of a courtyard in the temple complex appeared on the floor. Below the eaves on one side stood a LINGbot: squat, solid, topped by a spherical head. Instead of legs the robot roved about on wheels, and its hands were clasped together in front in meditation mudra, left hand on right, palms up, thumbs touching. The face, smooth as an egg, was devoid of features save for a tiny red light that flickered on the forehead.

The light turned green. She, having embodied the LINGbot, opened her eyes.

The shadow-draped trees rustled in the breeze, redolent of fragrant osmanthus. Hungrily, she drew a deep breath. Then she lifted her hands and moved her fingers one by one, luxuriating in the feeling of the air rushing between them.

A snail crawled slowly across the bare ground in the sun, leaving a dashed trail. Carefully, she caught it, picked it up, moved to the grass under a tree, and gently deposited the creature out of danger.

There were about twenty large water-filled copper basins in the courtyard, full of bobbing fish. The sun gleamed in the water, and fish playfully shuttled between the flickering rays. It was the third day of the Liberation Rite of Water and Land, and the most important rituals were about to take place. Starting at three in the morning, the sima-boundary barrier had been purified, and charms released to invite the attendance of various spirits. A flapping banner solemnly announced the initiation of the core mysteries.

In the afternoon, the ritual for releasing caged animals to gain merits would be performed in front of the main shrine hall. In order to prevent unscrupulous merchants from raising prices, the animals to be released had to be purchased by monks from various live markets half a month ahead of time.

She began her duties. First, she retrieved a few small basins and used a ladle to transfer the fish from the large basins to the small ones. A few of the fish were drifting upside-down, showing their white bellies. These she scooped out and transferred to a separate basin for observation. She felt sad for the dying fish, but then she remembered that the fish that would be successfully released later might not be considered fortunate either. It was likely that opportunistic fisherfolk were already waiting downstream by the river, hoping to catch the fish so that they could be resold as dinner.

Her mind drifted in memories of life and death. Twice a week she was allowed to perform tasks outside the sealed chamber by embodying a LINGbot. Most of the time, she went to hospitals, orphanages, retirement homes, animal shelters, mortuaries, and cemeteries. She had taken care of abandoned babies, abused pets, children with fatal illnesses, and elders on the verge of death. She had also taken care of the bodies of the dead, people as well as animals, chanting sutras for them, praying for their souls. She had adjusted to these duties more quickly than others had in similar circumstances, perhaps because she wasn't as sensitive to the impact of death, or perhaps because such tasks were the only way for her to experience the world outside the temple. The world was filled with disease, suffering, blood, death, and howls of mourning, but it was also filled with the smells and incandescence of life. Through the nimble hands of the LINGbot, she could touch life in its infinite variety, feel its fragility and strength, joy and pain, despair and hope.

Clacking footsteps entered the courtyard. A boy ran in, lively and rambunctious, about seven or eight years old. He wandered around the courtyard and knelt down at the rim of a large basin, sticking his hands into the water. The fish, frightened, darted away, and water splashed beyond the rim.

She wheeled herself over. "Please don't play with these fish. They'll be released later into the river."

The boy ignored her and continued his efforts, intent on catching a fish. She grabbed his hand, but the boy snatched it away. Enraged, he kicked her hard, and then grabbed a ladle to splash her with water.

She felt no pain. To protect embodied users, LINGbots often had their pain sensors adjusted to very high thresholds. Moreover, she was used to strangers or animals assaulting her when she was trying to carry out her duties in LINGbot form.

She scooped the boy up, holding him by the waist. The boy screamed and fought against her hold, but he couldn't overcome the pliant yet strong silicone arms. She stayed still, holding onto the boy, waiting for him to exhaust himself.

"This is a sacred place; please don't mar it with your disturbance."

She turned around and saw a woman dressed in white, a tiny red mole at the center of her forehead.

"Please put him down," the woman said.

She couldn't argue, so she set the boy down.

The woman bent down to the boy. "Why are you trying to catch a fish?"

The boy's face was scarlet red, but he refused to answer.

"If you really want one, then go ahead. But you may only catch one."

The boy immediately ran to the basin and bent over the rim. After a lot of splashing and churning, he finally managed to get his hands around a golden-red carp, as thick as his arm. He lifted the fish out of the water. The creature struggled hard, flapping and twisting. The boy laughed.

She was about to go over to save the fish, but the woman in white extended an index finger, gently tapped the head of the carp in the boy's arms, and then tapped the boy on the forehead. The boy shook, and suddenly his mouth gaped open as his tongue lolled about, a gurgling sound in his throat. Terror filled his eyes as his face turned a deep red like the liver of a pig. The carp dropped from his hands, snapping and hopping on the hard ground.

The woman in white picked up the fish and held it before the boy's face. "A fish cannot breathe outside water. If it's out of the water for too long, it will die. You can save it by returning it to where it's supposed to be."

Wide-eyed, the boy accepted the fish with trembling hands and deposited it back into the basin. The second the fish was back in the water, he managed to draw a deep breath, and his face gradually recovered its normal color.

"You should go back," said the woman. "Your mother has been looking for you."

The boy stood still for a moment and then burst out crying. Still sobbing, he ran for the courtyard exit.

The woman watched him depart and sighed. She turned to the LINGbot. "Did he hurt you?"

She shook her head.

"I've also embodied LINGbots to help out around the temple, and I've met some unreasonable visitors, too," the woman said. "You saved the fish's life; it's a merit."

Startled, she held still and then spoke for the first time. "I don't believe in karma."

"Then why did you try to save the fish?"

She tried to answer but couldn't find the words. Sunlight sparkled in the water. The carp, rescued from a catastrophe, joyously swung its tail in the tank, the sounds of splashing and sloshing echoing crisply from the walls.

The woman in white gazed at the fish. "Do you think fish experience pain?"

She hesitated. "I suppose they do."

"How do you know?"

"I . . . don't know."

The woman sighed. "The scientific community has debated the question of whether fish experience pain for decades. Some researchers discovered nociceptors in fish, sensors that send nervous impulses into the cognitive regions of the brain in a similar manner as corresponding pathways in higher vertebrates, not mere conditioned reflexes. But other researchers insist that the fish brain is too simple, devoid of the cortex found in primates or other mammals, so that fish cannot form thoughts that we should interpret as 'I am in pain.' Ultimately, because we're not fish, we cannot know if they experience joy or pain. Or, more precisely, we cannot know if the suffering of fish is comparable to the pain we experience as humans."

She listened, only half understanding. But something moved in her heart, as though a pebble had been dropped into a deep well, the reverberations indistinct and dark.

The woman took off a white ring from her right index finger and caressed it absentmindedly. After a moment, she added in a low voice, "Perhaps empathy exists not only between one human and another."

Curious, she asked, "What is that?"

"A gadget given to me by a friend. It's called LINGpain, capable of recording and replaying nociperception, the nervous system's responses to hurtful stimuli in different organisms. It allows a person to share pain with others, and to experience the pain of others. I hope that child can remember from now on the sacred nature of each life. Everyone suffers pain."

She pondered. Then she pressed her artificial hands together and bowed.

The woman bowed back. "I'll leave you to your work." She turned and left.

Silence returned to the courtyard. The fish continued their play in the sparkling water, as though unaware of all that had happened.

## Inviting Superior Beings and Making Offerings

THUNDER, LIGHTNING, RAIN SLAMMING AGAINST THE EAVES.

Inside, Xiao Wang sat across from an aged monk on putuan cushions. The aged monk was gaunt like a bamboo pole, his beard and brow snow-white, each hair erect like a silver needle.

Xiao Wang pressed her hands together and bowed. "The venerable monk is up late."

The aged monk replied, "Today is the ritual of puja worship. We have to invite the various Buddhas, bodhisattvas, and arhats of different degrees of enlightenment to attend the service and receive offerings. The first rite starts at three in the morning, so I'm up early, not staying up late."

"You're up to invite the bodhisattvas," Xiao Wang said, "whereas I've come without being invited."

"I never said you couldn't come," said the old monk.

"Have you had a chance to look over my reports?"

"I have read both. We've already made significant progress on integrating LINGcloud with LINGpain so that people can, through the medium of the cloud, experience one another's suffering. As for your suggestion to use LINGcloud to build schools in remote, impoverished regions, that is a deed of great merit. All the monks are very supportive of the idea. I've already asked those in Wenshu and Puxian Institutes to draft a plan of implementation as soon as possible. But everyone is so busy right now with the Liberation Rite, so I'm afraid you still have to wait a little longer"

Xiao Wang nodded. "I know you're busy, and I'm sorry to bother you in the middle of the night. But my heart is plagued with doubt, and I'm hoping you can help me by answering some questions."

"Go ahead."

"I checked the file for Householder Zhou. It turns out that his original name was Zhao Shizong."

The old monk said nothing.

Xiao Wang continued. "Eighteen years ago, Zhao Shizong's family died as the result of a horrible crime. While he was out of the country for business, the perpetrators broke into his home and tortured his wife, son, and daughter to death over a period of ten days. The crime wasn't discovered until half a month later, when neighbors reported the foul stench emanating from the apartment. The perpetrators took advantage of LINGmask software, which allowed users to change the appearance and voice of the subject of a video to simulate anyone else's, virtually undetectably. This was how they were able to pretend to be members of Zhao's family during video calls, making him think that nothing was wrong.

"That wasn't all. They recorded the entire process of their torture and murders, and then released the videos on the web. The videos were spliced together from footage taken by multiple homesecurity cameras and scenes shot from the perspectives of the victims and the perpetrators, as well as video chats between Zhao and the disguised perpetrators, synchronized to the tortures occurring at the time. The videos had been processed such that the faces of the perpetrators were replaced with blank masks, and viewers were encouraged to use LINGmask to replace the faces with their own or the faces of anyone else they liked.

"The vile videos spread like wildfire on the web, and many denounced the criminals and expressed their sympathy for the victims even as they pressed the download button. Some of the videos, after the substitution of other faces, would be re-uploaded. Although social media platforms tried hard to scrub the videos, they continued to go viral through encrypted and hidden channels."

The old monk sighed. "Amituofo. What a sin."

Xiao Wang went on. "At that time, someone revealed online that LINGmask had been developed by a group led by Zhao Shizong himself. Zhao's company had earlier found itself in controversy with the invention of LINGsee, a facial recognition–equipped mobile nanocamera platform that enabled the continuous tracking and filming of designated subjects. Many had accused Zhao's company of being careless with the resulting surveillance, privacy, and lack of consent issues.

"In response to the uproar, Zhao had led a small group to develop LINGmask, which could automatically seek out videos containing the face of the user and then replace it with a blank mask or another face. Some members of his group pointed out that the software would create even more safety issues and risked being abused by criminals, but the product was nonetheless brought to market and even achieved success.

"The history of LINGsee and LINGmask thus became a rallying point for some trolls, who attacked Zhao and used it as an excuse to deliberately spread the videos of his family's murders, claiming that Zhao was simply reaping what he had sown, that their deaths were a manifestation of karma."

The old monk shook his head. "To deliberately detract from the truth with falsehoods; to confuse good and evil. What a sin."

Xiao Wang continued. "Eighteen months later, the police finally caught the murderers. They were all youths: three boys and a girl. The youngest, the girl, was only twelve, and the oldest, a boy, had just turned eighteen. Due to the heinous nature of the crime, the three boys were sentenced to terms of imprisonment, but the girl, who had not reached the age of legal responsibility, had to be released without penalty. The girl and her mother moved several times, but each time a neighbor would leak her information to the media and reporters would follow them around, giving them no chance to live a normal life. Yet, half a year later, the mother and daughter simply vanished and were never heard from again."

The old monk said nothing. The sound of rain filled the silence.

Xiao Wang went on. "At first, I thought the two of them, like Zhao Shizong, had adopted new identities and emigrated abroad. But just now, I realized that Zhao Shizong must have been trying to track them down all these years. His sudden return to China and visit to Lingyin Temple must have something to do with the pair. Am I right?"

The old monk said nothing, but held out a hand, palm up. A dharma cloud swirled above his palm, gradually coalescing into a projection of a woman sitting alone chanting sutras, her hair draping to the floor.

Xiao Wang gasped. "Is she ..."

The dharma cloud shifted and coalesced into a projection of the scene from noon a day earlier, when Xiao Wang had encountered the LINGbot in that courtyard.

"So it is her! Has she been at Lingyin Temple all these years?"

The old monk waved away the projection, pressed his palms together, and bowed. "Amituofo. You guessed right."

"But how ...? Considering how securely Lingyin Temple protects its data, I cannot imagine how Zhao Shizong would find her. Did ... did you ..."

The old monk said nothing.

"You invited him to come." Xiao Wang paused. "I suppose you wish to resolve this sin-steeped enmity?"

"Whether it can be resolved is not up to me, but to the two of them and their fate."

"But I also found out something about those three boys. After their release from prison, they all soon disappeared, and no corpses were ever found. The police suspected that Zhao Shizong was behind it somehow, but there was insufficient evidence even to initiate a formal investigation."

The old monk shook his head. "Vengeance begets vengeance; retribution multiplies upon retribution. Where is the end to the chain of suffering? What a sin."

"He really was behind those disappearances, then? Even though you know his intent, you're willing to take this risk?"

"The sea of suffering is boundless; only by turning back can you find the shore."

"But one who is lost in the sea of suffering cannot turn back! How can eighteen years of bitter resentment be dissolved with some recited scriptures?"

The old monk said nothing.

The temple clappers clacked outside. The rain had stopped.

At length, Xiao Wang said, "I still have one more question that I've been meaning to ask for a long time."

The old monk said nothing.

"In your room, there is a memorial tablet with no name of the deceased. Whose soul are you trying to relieve from suffering?"

"It's late. Let's talk about this another time."

Xiao Wang sighed. Her projection dissolved into dharma cloud and dissipated.

## Bestowing upon the Dead the Bodhisattva Precepts

IT WAS THE FIFTH DAY OF THE LIBERATION RITE, WHEN THE MONKS WOULD INVITE the Inferior Beings, meaning all sentient beings in the six destinies and lonely ghosts and lost souls, to receive relief.

At dawn, the monks issued the plea for the guardian spirits to release all sentient beings in the six destinies from their fetters. From noon to evening, a feast of fourteen tables was laid out for the sentient beings, so that they might bathe and change, find their path, resolve their hatred and enmity, cleanse their three unwholesome roots and six karmic courses. That night, the monks would bestow upon the dead the bodhisattva precepts, guide them to repent for their sinful deeds, plant the desire for enlightenment in their hearts, and help them receive the great vehicle precepts, committing to doing good to amend for past evil.

Was it really possible to repent, to receive the precepts, and to be absolved from the sea of suffering and begin a new life?

Xiao Wang stood before the window, watching the waning crescent moon low in the sky.

As a little girl, she had listened to the teachings of the master monks in the temple: the wandering wheel of samsara; the retribution and rewards of karma; a life of good leading to a next life of wealth and happiness; a life of evil leading to a next life of suffering as a beast of burden. But she thought those were mere fairy tales made up by adults to scare children into behaving.

As she grew up, gradually she began to understand that everything in the universe *was* connected. Every word, every act, every emotional outburst or passing fancy generated consequence upon consequence, links in an ending chain. A piece of trash carelessly discarded today would eventually return in the form of polluted air and water to redound upon the original disposer. A hurtful word aimed at a child in a moment of rage today could lead to a bitter grown-up committing murder in the future. To see karma in action, one need not wait for reincarnation.

This was even more so in an increasingly complex, mutually dependent technological world. Each individual's understanding of reality in an overwhelming tsunami of data was so limited, so fragmentary. To consume a piece of steak, to take a sip of milk, to buy a new pair of jeans, to upgrade to a new phone—all these actions meant the suffering of people and animals. You were oblivious, immersed in the pleasure of consumption, with colorful advertisements and fancy packaging isolating the pain-wracked bodies from your view. To maintain equanimity, you relied on malicious labels placed on those who differed from you by race, gender, class, or culture, hoping that they would go as far away as possible, or even better, die off, without ever acknowledging the connection between everything you possessed and everything they couldn't possess. Everyone's fate was intimately entwined with the fate of everyone else, but selfish desires made it impossible for anyone to feel or imagine the suffering of others. Everyone had their eyes clouded by the tiny slice of selective data they had access to and the illusions woven by the media, unable to perceive the whole, committing error upon error. Everyone complained that the morality of the world was in decline, but refused to see their own responsibility, avoided the need to change...

Xiao Wang recalled reading the Buddhist scripture passage explaining that "ignorance" was the first of the twelve nidanas that together formed the chain of dependent phenomena, the basis of all suffering and rebirth.

She had no idea then what this "ignorance" meant. But now, she realized, the willful unnoticing, unknowing, unseeing, unfeeling of modernity—this was true *ignorance*.

She clasped her hands together in lotus mudra, gazing down at the 卍 in the cupped palms, leaping like a ghostly flame.

Was it really possible to break out of this state with the aid of technology?

Twelve years ago, the launch of the 卍 system had generated much controversy and skepticism. The system relied on big data and pattern recognition to track and record every act by every person in every moment, from an angry word to a murderous lunge, and kept a ledger of credits and debits composed of good deeds and evil, from which it computed the chain of karmic consequence over time. You had the right to access your own 卍 account, but that was all. It was impossible to compare your own progress with that of anyone else, and the data was inaccessible in any other way. You needed not worry about law enforcement knocking down your door based on 卍 records, and you could be free from the anxiety that judges in the afterlife would cite these computerized records in the trial before rebirth. But deep at night, as you sat alone in your room, the 卍 records might bring you a pang of regret, a shiver of distress.

Oddly, though the vast majority of the population would never come to Lingyin Temple even once to view their own 卍 account, talk about it became the new fashion. Memes and posts teaching people how to extinguish karmic hindrances and accumulate benefits spread virally: vegetarianism, chanting sutras, Chan meditation, spiritual cultivation, abstention from tobacco and alcohol, offering incenses and prayers, donating to temples...

She had always remained skeptical. Such explicit attempts at manipulating the ledger clearly deviated from the original intent; in fact, the 卍 system had mutated into a performance, a game, perhaps a business.

But then, who could really tell the system designer's original intent? If those who had done evil donated their ill-begotten gains to the temple to accumulate merits, and the temple then used the funds to support acts of charity, wasn't a kind of karmic balance achieved?

Lingyin Temple, like the Venerable Zhengxuan, was shrouded in layers of enigma and mystery, impossible to see clearly.

Tomorrow was the day those two would meet.

She wasn't a devout Buddhist, but nonetheless, she pressed her palms together and prayed for them.

# Fulfillment of Hunger

ONCE AGAIN, HE WAS STARTLED AWAKE FROM HIS NIGHTMARE.

The nightmare was likely to be with him for the rest of his life.

It was the path he had chosen. Even if he were to be plunged into an ocean of fire and dragged across a mountain of knives—to be trapped in Avici Hell for eternity—he had to keep on going; he could not turn back.

He held his hands together and cracked his knuckles, focusing on the faint, sharp snaps.

It was time.

SHE KNEW IT WAS TIME.

By her own calculation, she had been inside this sealed chamber for a full sixteen years. From a child, she had progressed into middle age.

The door to the outside was open. She had dreamed of this moment countless times, but she had never expected to feel so much trepidation when the time came.

She got up from her putuan. Clutching the prayer beads, she stepped outside the door, into the long, dark corridor.

There was no one inside.

FOR A LONG TIME, HE WALKED BY HIMSELF THROUGH THE DARKNESS, UNTIL HE SAW a glint of light. As he approached, he realized that the glint was from a giant 卍, like a flaming lotus towering in the middle of the road.

He extended a hand to touch it, and on contact the character burst into a thousand thousand red and blue points of light, like a scattering of seeds sprouting into entwining vines, weaving into a complex graphic. Three bright red flames stood out, like glowing embers or pulsing tumors. Bloody light poured over him, enclosing him in a tight, seamless cocoon.

Through the red light he saw the faces of three young men. He had been so patient over the years, carefully laying out traps to lure them in, and then kidnapping them, imprisoning them, torturing them, killing them, destroying the bodies, and erasing the evidence. . . . He wanted them to experience all the pain they had inflicted on his family. An eye for an eye, a tooth for a tooth. Killing could not free him from his own nightmare, but it gave him a reason to keep on living. There was no other choice. None.

There was one more. Just one more.

Through the curtain of blue and red light, he saw a woman's face.

She walked toward her own 卍, until her whole body was immersed in it, like returning to the beginning of her life.

A voice began to murmur in her ear, explaining to her the chain of causes and consequences.

*Your parents were introduced to each other, and both sets of grandparents wanted them to marry. After a half-year they rushed into marriage, a union troubled from the start. Your father had a bad temper, drank excessively, beat your mother. Your mother wanted a divorce, but her own parents forced her to return to her husband again and again. All her relatives and friends counseled patience, telling her that she needed to have a child, that everything would be better with a child. During her pregnancy, her husband almost drowned her in the bathtub. She survived and gave birth to you.*

*You didn't have a bright childhood. Your father hit your mother, and your mother hit you in turn. One day, you stumbled into an encrypted directory your mother had concealed on the computer and managed to open it. The directory was filled with detailed notes on how to kill someone, gleaned from crime dramas and the news, with the videos carefully annotated and categorized. You were so intrigued that you went through these plans every chance you got and practiced some of them on stray dogs and cats and neighbors' pets. These notes became the most influential text of your childhood.*

*You discovered your father's secret as well. He liked to covertly record his sexual encounters and share the videos with a close circle of friends. You gained access to these recordings and watched them, learning about sex through the grimaces and moans of these unknown women. Later, you learned to use credentials stolen from your father to enter hidden corners of the web, where you used your father's videos to buy favors and gifts from boys and men. For the first time in your life, you learned how it felt to have power.*

*You found a group of teenaged boys to threaten your mother, so that she would stop beating you. You learned to hide at the homes of your classmates when your father was home. You observed your friends' families carefully, ferreting out their secrets. You were sure that every family concealed darkness and ugliness, and the joy they showed others was an illusion. You learned to use miniature hidden cameras to spy on these people.*

*The parents of one boy discovered his stash of sex videos, and he confessed that he had obtained them from you. His parents contacted the families of other classmates and discovered the extent of your little pornographic club. They went to the school authorities and demanded your expulsion so that their pure little angels could be saved from your corruption. You and your mother were summoned to the principal's office, where dozens of adults surrounded you, screaming, cursing, calling you bitch, whore, c\*\*\*. Your mother knelt on the floor in silence, enduring the onslaught the way she would endure a beating from your father.*

*You could no longer go to school. You spent your days online, and soon learned how to navigate the dark web, exploring an even more grotesque, twisted world inhabited by demons and monsters. With the aid of LINGmask, you could make yourself the protagonist of videos shot by murderers, but the satisfaction soon wore off. You needed more.*

*You made a plan. You found three boys willing to help. They wanted to have some fun by shooting videos that would arouse their numbed senses. You wanted to experience the reality of how it felt to kill.*

*For your target, you picked the family of the boy who had told on you. It wasn't an act of vengeance; it was because you were familiar with the home's layout.*

*All those classmates, their parents, the strangers on the web who swapped sex videos, your parents, their parents, the people who introduced them, counseled them to remain married and have a child—all of them planted the seeds of evil, committed acts that would blossom into sin.*

*Your co-conspirators were imprisoned, but you were released. Your father had gone missing, and your mother took you and moved again and again. But no matter where you went, someone would betray you to the media, to the neighborhood, letting everyone know that you were a cold-blooded murderer. Schools refused to take you. Neighbors gathered in front of your door demanding that you leave. Reporters followed you around with their cameras, sticking bundles of cash in your mother's hand so that they could buy the rights to your story. Your mother locked you inside the home, a prison by another name.*

*You remember that cold snowy night. Your mother knelt outside the gates of Lingyin Temple and touched her forehead to the ground again and again. A leash was tied to her wrist, and the other end was tied to your ankle. Hungry, cold, you fell asleep, until you were awakened by the tolling of the temple bells.*

THE BLUE AND RED LIGHTS FADED LIKE STREAMING SAND. FINALLY, HE COULD SEE that face clearly. There was a scar on her lip, pulling up the right corner of her mouth. The left corner, on the other hand, drooped, leaving the impression of an eerie smirk.

That face, that smirk. He would never forget it.

He saw the features on that face twisting in terror. He saw the lips trembling without making a noise. He saw the woman stumbling away, turning to escape into the darkness.

Blood surged in his chest; fingernails bit into his palm. He pursued.

•••

**TWO SETS OF FOOTSTEPS BROKE THE SILENCE IN THE CORRIDOR.**

She ran and ran. She ran into a temple hall immersed in darkness, with only a single oil lamp lit before the statue of the Buddha.

A hand seized her hair. A heavy body slammed her to the ground. She struggled, biting and scratching and kicking and punching like a cornered animal. With a grunt, her captor fell to the ground. She leaped onto him and wrapped her prayer beads around his neck, tightening the noose with all her strength.

The oil lamp's light flickered. She saw that his face was turning red and then purple. His eyes bulged out of their sockets. The veins in his temples swelled and pulsed. She could feel strength draining out of her arms. But he was still alive. She could still hear gasps and gurgles from his throat.

She grabbed a putuan nearby and pressed it against his face. She pressed all her weight against it. Time crawled by slowly. His chest no longer heaved, though his hands continued to spasm, striking the hard ground again and again, like fish out of water.

She tossed away the putuan, panting from the exertion, looking around for another weapon. She saw a heavy brazier on the altar. Slowly, she crawled over and retrieved it. After a moment of hesitation, she raised it above her head and smashed it down on his head.

Once. Twice. Thrice...

**HE DIDN'T KNOW HOW MUCH TIME HAD PASSED.**

He killed her, over and over, replaying scenes that he had imagined countless times. She was no longer recognizable; every inch of her body was covered in blood and gore. Yet, she remained alive.

He sat on the ground, leaning against a pillar for support. His body felt as though it were made of mud. Dragging her broken body, she crawled toward him through the blood smeared across the ground, one inch, another inch, another, another...

His nightmare had come true. He was trapped in Avici Hell.

Finally, she had crawled in front of him, and one hand grabbed him by the knee, a bloody, torn hand. Every fingernail had been plucked out; every joint was bent at an unnatural angle.

With his last ounce of strength, he pressed a palm against her forehead, keeping her from getting any closer. She calmed down and wrapped her arms around his knees, curling upon on the ground in the fetal position.

He suddenly remembered how, years ago, his little daughter had liked to listen to him tell stories in this position. He would caress her forehead, to smooth out the hairs moistened by her sweat.

He broke down and howled in grief.

SHE REMEMBERED HOW, AS A CHILD, SHE SOMETIMES LIKED TO STEAL INTO HER parents' bed deep at night and nestle between their slumbering bodies, pulling their arms over herself. Amid her parents' thunderous snores, she felt at peace and safe, ready to fall asleep herself. But she never dared to really drift off. As soon as one of her parents turned in sleep, she would leap to a corner of the bed, ready to conceal herself under the bed if necessary.

For so many years, no one had wrapped their arms around her like that.

She reclined in his embrace and placed his unmoving arms around her body. His blood soaked into her, but she could still feel a trace of warmth, seeping from his chest into her back, spreading to her whole body.

HE STRAIGHTENED HER TWISTED JOINTS AND CAREFULLY PIECED THE TORN FLESH back together, smoothing the mangled skin into place. He tore off pieces of his monk's robe, dipped the cloth in the holy nectar kept in front of the Buddha, and wiped away the blood from her face. He washed her face, brushed her hair, took off his own monk's robe, and put it on her.

He arranged her body in the pose of Guanyin, the Bodhisattva of Compassion. He took three steps back and knelt to pray.

SHE HUNG THE PRAYER BEADS THAT HAD KEPT HER COMPANY FOR SIXTEEN YEARS around his neck. She pressed her hands together, bowed, and began to recite the *Repentance of the Emperor of Liang*.

> The cleansing dew from willow branches
> Sprinkles to every corner of the three thousand great thousand worlds.
> Empty nature, eight virtues, beneficence to the six destinies.
> Let ghosts be released from their needle-swallowing hunger; let sins be unknotted and
> transgressions dissolved; let the flames of hell be turned into a red lotus.
> Namo Pure and Serene Bodhisattva, Mahasattva.

Melodious voices sounded in the air, coalescing into falling blossoms.

They opened their eyes simultaneously. They saw each other, as well as the self in the eyes of the other.

They had been sitting thus, face to face, since the beginning of the ceremony. The dharma cloud created illusions that enveloped them, and also connected them by eye, ear, nose, tongue, body, mind, heart, so that they could feel each other's pain and suffering, crime and punishment, good and evil, love and hate, causes and consequences.

The dharma cloud dissipated, revealing the brightly lit Hall of the Medicine Buddha. Inside, the final ritual for feeding the hungry ghosts was at its climax. In the hall, the Venerable Zhengxuan, bell and incense in hand, was inviting Ksitigarbha, Bodhisattva of the Hell Realm, to lead the souls of the participants' loved ones and the souls of the lost to the rite.

Two people sat, face to face, in the hall.

"We invite, with steady hearts, the lost souls of the stubborn and persistent, deprived of the knowledge of the Buddha: the uncivilized and barbaric, the blind and deaf, the hardworking servants laid low, the maids injured by jealousy. To scorn the three treasures is to accumulate sin like the grains of sand on a river shore; to dishonor parents is to fill the cosmos with offense. Alas! When will the sun rise on the endless night? When will spring be known in the lightless land?"

Bells tolled and incense smoke swirled, connecting the living with the dead.

## Farewell to the Enlightened

IT WAS THE SEVENTH DAY OF THE LIBERATION RITE, WHEN THE MERITORIOUS ACTS would come to fruition.

In the morning, the monks prepared a feast of delicious treats as offerings to the enlightened beings who had come to attend the Liberation Rite. Incense was lit to express the wish that sentient beings would thus depart from the sea of suffering and reach the Western Pureland of Ultimate Bliss.

The faithful took down the memorial tablets for the deceased and brought them outside the ritual hall. The temple had a spirit wall, yellow in wash, topped by black tiles, with a sign declaring, "Western Pureland Is but a Foot Away." A giant boat made of paper sat on the plaza in front of the spirit wall, and all the memorial tablets were placed on the boat. The venerable monks began to chant, bidding farewell to the various bodhisattvas as they ascended the cloud paths and to the sentient beings of the six destinies as they began the journey to the Pureland. Amid chants of *Amituofo*, firecrackers were set off, and the paper boat was set aflame. As the fire roared, everything mundane turned to swirling ashes heading for the horizon.

THAT EVENING, THE AGED MONK AND XIAO WANG WALKED SIDE BY SIDE TO THE western gate of the temple.

Xiao Wang pressed her hands together and bowed. "You've walked me far enough. Thank you."

"Take care on your journey."

"The weather is turning cold, be careful that you don't catch a chill."

"You as well."

Two birds hopped on a bough overhead, chirping one after the other as though engaged in some singing contest. Under the tree, the two humans remained standing.

"The sin-steeped enmity has been resolved," said Xiao Wang. "You've accomplished what you set out to do."

"It's only a start."

Xiao Wang sighed. "To construct a system that would allow all to share one another's pain . . . that is a task infinitely more complex and harder than the 业 system. Since you'll be retiring at the end of the year, perhaps it's best to pass the task onto your successor."

The old monk nodded. "You're right."

"I noticed," said Xiao Wang, "that earlier you had placed the nameless memorial tablet on the boat to the Western Pureland."

"Every year the nameless tablet is burned," said the old monk. "I'll make a new one and keep it till next year."

"Then . . . the nameless tablet wasn't for that pair?"

The old monk was silent for a while. "The tablets are for the souls of those who have not yet died, but are destined to die."

"What do you mean, exactly?"

"Years ago, I pressed ahead with the LINGcart project. To satisfy the transportation needs of the tens of millions in a city with a network of hundreds of thousands of spherical carts traveling in tubes required the development of an extremely complicated set of algorithms. During simulations, I realized that a most difficult problem was how the system would respond to disasters such as earthquakes, fires, terrorist attacks. However, no matter how I piled safeguards on redundancies, added improvements upon optimizations, I couldn't eliminate situations in which the system had to make a choice: a choice to save the many by sacrificing the few."

He held out a hand, and a projection appeared above his palm: a spiderweb of intersecting rails, with countless green beads streaming along. Abruptly, an expanding red section appeared in the middle. The green beads changed their trajectories to flee the danger. After ten seconds, the vast majority of the green beads managed to escape, but a few that couldn't get out in time turned red.

"At one time, proponents of driverless cars argued that the more self-driving cars were in the streets, the fewer accidents caused by human frailty. Although driverless cars were not accident-free, it was an inevitable cost of technological progress, the sacrifice of one to save the lives of a thousand thousand. At one time I agreed with their logic, but I failed to understand that for the few who died as sacrifices, they were not mere data, but beings of flesh and blood, who could cry and suffer pain, who had loved ones waiting in vain for their return."

The projection above his palm turned into blurred footage from a surveillance camera: in the dark, a driverless car swerved to the right to avoid a school bus that had run through a red light; the driverless car was headed for a pedestrian standing at the side of the road. The video froze. It was impossible to make out the face of the lone pedestrian, overexposed in the headlights.

He extended his other hand, as though trying to shield the tiny figure from the oncoming, unstoppable vehicle. The dharma cloud projection trembled between his fingers.

"Should the one die, or the hundred? To even ask such a question was to already have set in motion the chains of karma. I once believed that in order to escape from this dilemma, the only solution was LINGcart. Only too late did I realize that the same ethical trap waited for me there. I tried to convince myself: the system was extremely safe, with vanishingly small probabilities for accidents. Even if the unexpected occurred, the system would make the most rational choice to save the most lives by sacrificing the smallest possible number. But I couldn't tolerate treating those who would be killed as mere data. The Buddha said that in the endless cycles of samsara, everyone would have a chance to be the parent of everyone else. They were no different from those related to me by blood, bound by links of love and pain."

The projection above his palm turned into a family portrait: father, mother, son, daughter. They sat around a dinner table, looking joyous.

"This nameless tablet is for all those destined to die in my simulations. To date, LINGcart has never had such an accident, and so they remain alive. But they are fated to be sacrificed one day. They are lost souls wandering the desolation of my algorithms, waiting for a chance to leap out to consume the living. Every day, I chant and pray, hoping that they would reach the Western Pureland of Ultimate Bliss instead of disturbing the living. I try to remind myself: the seeds of sin and evil are planted in a single thought. I can only make amends by bending my heart toward good through faith."

Xiao Wang let out a held breath and waved her hand to enlarge the projection. Her fingers slowly brushed across the four faces: father, mother, big brother, little sister. There was a red mole at the center of little sister's forehead, pure vermilion.

Tears glinted in her eyes. Then she laughed. "This morning, I dreamed of Mom."

The old monk said nothing.

"I've never believed that the dead could visit the living in dreams," said Xiao Wang. "But there is something special about this dream. She was sitting by my bed, her hand on my head, telling me that she was so glad to see me grown up, looking so different, except for the mole on my forehead. She also said that the Liberation Rite of Water and Land accumulated so much merit that she also benefited. I suppose she must have known that I came to visit you, and so she came to join our reunion."

The old monk waved away the dharma cloud projection. "It has been many years since I dreamed of her."

After a beat, Xiao Wang said, "I brought you a present."

She opened her palm. The projection of a little girl took form. She was dressed in a flowery dress, with two braids. A pair of dark, shiny eyes sparkled in a sun-darkened face. A wisp of LINGcloud turned into a hovering piano keyboard, and the girl began to play. "Ode to Joy" filled the air.

"Her name is Qianqian," said Xiao Wang. "She was in the class I taught in Baizhu Village in Yunnan. I asked her to play this for you."

The girl finished. She looked into the camera and smiled shyly. "Thank you."

Xian Wang's voice came from beyond the frame. "Who are you thanking?"

"I want to say thank you to Grandpa Monk."

A smile relaxed the aged monk's wrinkled face. He pressed his hands together. "Amituofo. What a wonder."

Xiao Wang smiled in response. She also pressed her hands together. "I'm leaving."

She stepped over the threshold and walked away along Tianzhu Path.

The sun, filtered by the tree branches, dappled the road. Birds hidden in the bamboo groves lining the path sang and chirped, as though wishing her a pleasant journey.

[*Translator's note:* I'm grateful to Dr. Kate Lingley for her advice in the translation of Buddhist concepts and terms. Any errors are entirely mine.]

# ARTWORK: TATIANA PLAKHOVA

AS A LITTLE GIRL, TATIANA PLAKHOVA USED TO PLAY A GAME. HER ROOM WAS A spaceship, with the window as the screen of the cockpit or bridge. She traveled around the universe visiting new and fascinating worlds. Tatiana graduated from Moscow State University with a master's in social psychology, and then studied at the Higher Academic School of Graphic Design. She now works as an art director, graphic designer, and illustrator. Today she likes to dream about living, breathing "big data" networks and imagine how they might grow and evolve.

In her data visualization work for an international roster of clients, Tatiana tries to make the data more engaging and beautiful, to speak a more immediate language to the viewer. In her own work, she wants to show a new way of producing and seeing "infographic" drawings. "We should change the way we see this form of art. It could be much more inspiring, a way for us to connect," Tatiana says. "Everything we see is biological, mathematical, or geological information. It can also be cultural patterns or any other thing. Complexity Graphics works are based on mathematical simplicity and harmony. I would describe them as infographic abstracts. This mathematical style helps me to illustrate everything from biological cells to deep space and meditative worlds. That's why I admire math, because it's everywhere and nowhere."

Tatiana's work, which is produced with mixed media software, bridges the often invisible worlds of music, scientific information and data, and the visual beauty of art. Her images allude to the invisible webs and relationships between people, landscapes, and worlds beyond this one. "Everything is connection, everything is information."

Tatiana lives in Moscow, and her work can be found at complexity graphics.com.

# CONTRIBUTORS

**James Patrick Kelly** has won the Hugo, Nebula, and Locus awards. His most recent books are "King of the Dogs, Queen of the Cats" (2020), a novella; *The Promise of Space and Other Stories* (2018), a short story collection; and *Mother Go* (2017), a novel. 2016 saw the publication of a career retrospective *Masters of Science Fiction: James Patrick Kelly*. His fiction has been translated into eighteen languages. With John Kessel, he has co-edited five anthologies. Jim writes a column on the internet for *Asimov's Science Fiction* magazine. Find him at www.jimkelly.net.

**Mary Robinette Kowal** is the author of the Lady Astronaut duology and historical fantasy novels: The Glamourist Histories series and Ghost Talkers. She's a member of the award-winning podcast *Writing Excuses* and has received the Astounding Award for Best New Writer, four Hugo awards, the RT Reviews award for Best Fantasy Novel, the Nebula, and Locus awards. Stories have appeared in *Asimov's Science Fiction* magazine, *Strange Horizons*, several Year's Best anthologies, and her collections *Word Puppets* and *Scenting the Dark and Other Stories*. Her novel *Calculating Stars* won the Hugo, Nebula, and Locus awards. As a professional puppeteer and voice actor (SAG/AFTRA), Mary Robinette has performed for *LazyTown* (CBS), the Center for Puppetry Arts, and Jim Henson Pictures, and she founded Other Hand Productions. Her designs have garnered two UNIMA-USA Citations of Excellence, the highest award an American puppeteer can achieve. She records fiction for authors such as Seanan McGuire, Cory Doctorow, and John Scalzi. Mary Robinette lives in Nashville with her husband Rob and over a dozen manual typewriters. Visit her at maryrobinettekowal.com.

**Nancy Kress** (nancykress.com) is the author of thirty-four books, including twenty-six novels, four collections of short stories, and three books on writing. Her work has won six Nebulas, two Hugos, a Sturgeon, and the John W. Campbell Memorial Award. Her most recent work is *Terran Tomorrow* (2018), the final book in her Yesterday's Kin trilogy. Nancy's fiction has been translated into Swedish, Danish, French, Italian, German, Spanish, Polish, Croatian, Chinese, Lithuanian, Romanian, Japanese, Korean, Hebrew, Russian, and Klingon, none of which she can read. In addition to writing, Nancy often teaches at various venues around the country and abroad, including a visiting lectureship at the University of Leipzig, a 2017 writing class in Beijing, and the annual intensive workshop TaosToolbox. Nancy lives in Seattle with her husband, writer Jack Skillingstead.

Rich Larson was born in Galmi, Niger, and has lived in Canada, the United States, and Spain. He is now based in Prague, Czech Republic. Rich is the author of *Annex* and *Cypher*, as well as over a hundred short stories—some of the best of which can be found in his collection *Tomorrow Factory*. His work has been translated into Polish, Czech, French, Italian, Vietnamese, and Chinese. Find him at patreon.com/richlarson.

Ken Liu (http://kenliu.name) is an American author of speculative fiction. A winner of the Nebula, Hugo, and World Fantasy awards, he wrote The Dandelion Dynasty, a silk-punk epic fantasy series (starting with *The Grace of Kings*), and the short story collections *The Paper Menagerie and Other Stories* and *The Hidden Girl and Other Stories*. He also authored the novel *Star Wars: The Legends of Luke Skywalker*. Prior to becoming a full-time writer, Ken worked as a software engineer, corporate lawyer, and litigation consultant. Ken frequently speaks at conferences and universities on a variety of topics, including futurism, cryptocurrency, history of technology, bookmaking, the mathematics of origami, and other subjects of his expertise. Ken is also the translator for Liu Cixin's *The Three-Body Problem*, Hao Jingfang's "Folding Beijing," and Chen Qiufan's *Waste Tide*, and he is the editor of *Invisible Planets and Broken Stars*, anthologies of contemporary Chinese science fiction. He lives with his family near Boston, Massachusetts.

Sam J. Miller is the Nebula Award–winning author of *The Art of Starving* (an NPR best of the year) and *Blackfish City* (a best book of the year for *Vulture, The Washington Post*, Barnes & Noble, and more—and a "Must Read" in *Entertainment Weekly* and *O: The Oprah Winfrey Magazine*). A recipient of the Shirley Jackson Award and the John W. Campbell Memorial Award, and a graduate of the Clarion Writers' Workshop, Sam's short stories have been nominated for the World Fantasy, Theodore Sturgeon, and Locus awards and reprinted in dozens of anthologies. He lives in New York City and at samjmiller.com.

Annalee Newitz (techsploitation.com) writes science fiction and nonfiction. They are the author of the novel *Autonomous*, nominated for the Nebula and Locus awards, and winner of the Lambda Literary Award. As a science journalist, they are a contributing opinion writer for the *New York Times* and have a monthly column in *New Scientist*. Annalee has published in the *Washington Post, Slate, Popular Science, Ars Technica*, the *New Yorker*, and the *Atlantic*, among others. They are also the co-host of the podcast *Our Opinions Are Correct*. They were the founder of *io9*, and served as the editor-in-chief of *Gizmodo*. Their new novel, *The Future of Another Timeline*, was published in September 2019.

Suzanne Palmer (zanzjan.net) is a Linux and database system administrator for the sciences at Smith College by day, and an author of science fiction and fantasy by night. Her work appears regularly in the pages of *Asimov's Science Fiction* magazine and *Clarkesworld*, and her story "The Secret Life of Bots" won the 2018 Hugo Award for Best Novelette. Her first novel, *Finder* (2018), was the first book in the Finder Chronicles series. The second book in the series, *Driving the Deep*, will be out in May 2020.

**Tatiana Plakhova** is an art director, graphic designer, and illustrator living in Moscow, Russia. Outlets that have featured her work include the *Wall Street Journal*, the *Harvard Business Review*, *New Scientist*, *Nature*, *WIRED* magazine, and HarperCollins Publishers. Among her present and past clients are Paramount Pictures, Sony Music, the 54th Annual Grammy Awards, L'Oréal Paris, Breakthrough Initiatives, and HP.

**Cadwell Turnbull** (cadwellturnbull.com) is the author of *The Lesson*. He received an MFA in creative writing and an MA in linguistics at North Carolina State University, where he was the winner of the 2014 NCSU Prize for Short Fiction for his short story "Ears." Cadwell also attended Clarion West 2016. His short stories have appeared in *Asimov's Science Fiction* magazine, *Lightspeed, Nightmare*, and *The Verge*. His short story "Loneliness Is in Your Blood" was selected for *The Best American Science Fiction and Fantasy 2018*. His novelette "Other Worlds and This One" was also selected as a notable story in the anthology. His short story "Jump" was selected for *The Year's Best Science Fiction and Fantasy 2019*. Cadwell lives in Somerville, Massachusetts.

**Nick Wolven**'s short science fiction has appeared in *Analog, Asimov's Science Fiction* magazine, *Clarkesworld, The Magazine of Fantasy & Science Fiction, The New England Review*, and *WIRED* magazine. Nick's stories often focus on the unintended social consequences of rapid technological advance. He lives in the Bronx and is an inveterate recluse; he is not active online, but can be reached at nick.wolven@gmail.com.

**Xia Jia** is the pen name of Wang Yao, an associate professor of Chinese Literature at Xi'an Jiaotong University and a visiting scholar at University of California, Riverside, from 2018 to 2019. Her academic collection on contemporary Chinese science fiction *Coordinates of the Future* was published in 2019. She has been publishing speculative fiction since college. Seven of her stories have won the Galaxy Award, China's most prestigious science fiction award. So far she has published a fantasy novel *Odyssey of China Fantasy: On the Road* (2010), as well as three science fiction collections: *The Demon Enslaving Flask* (2012), *A Time Beyond Your Reach* (2017), and *Xi'an City Is Falling Down* (2018). In English translation, she has been published in *Clarkesworld* and other venues. Her first story written in English, "Let's Have a Talk," was published in *Nature* in 2015. Her first English collection, *A Summer Beyond Your Reach: Stories*, will be published in 2020. She is also engaged in other science fiction related works, including academic research, translation, screenwriting, editing, and teaching creative writing.

**Lisa Yaszek** is a professor in the School of Literature, Media, and Communication at Georgia Tech, where she explores science fiction as a global language crossing centuries, continents, and cultures. Lisa's books include *Galactic Suburbia: Recovering Women's Science Fiction* (2008); *Sisters of Tomorrow: The First Women of Science Fiction* (2016); and *The Future Is Female! 25 Classic Science Fiction Stories by Women* (2018). Her ideas have been featured in the *Washington Post, Food and Wine Magazine*, and *USA Today*, and she has been an expert commentator for the BBC4's *Stranger Than Sci Fi*, WIRED. com's *Geek's Guide to the Galaxy*, and the AMC miniseries *James Cameron's Story of Science Fiction*. A past president of the Science Fiction Research Association, Lisa currently serves as a juror for the John W. Campbell Memorial Award and the Eugie Foster Memorial Award for Short Fiction.

**Sheila Williams** is the multiple Hugo Award–winning editor of *Asimov's Science Fiction* magazine. She is also the editor or co-editor of over two dozen anthologies. Sheila writes regular essays and editorial pieces, and she is the co-founder of the Dell Magazines Award for Undergraduate Excellence in Science Fiction and Fantasy Short Story Writing, which is given out each year in Orlando, Florida, by the International Association for the Fantastic in the Arts. She lives with her family in New York City.